future arquitecturas s.l. 编

万象建筑新闻 3
panorama architecture newspaper 3

图书在版编目（CIP）数据

万象建筑新闻.3/《未来建筑》杂志社编. — 天津：
天津大学出版社，2012.6
ISBN 978-7-5618-4386-4

Ⅰ.①万… Ⅱ.①未… Ⅲ.①建筑设计 Ⅳ.①TU2

中国版本图书馆CIP数据核字（2012）第130516号

责任编辑　朱玉红

出版发行　天津大学出版社
出 版 人　杨欢
地　　址　天津市卫津路92号天津大学内（邮编：300072）
电　　话　发行部：022—27403647　邮购部：022—27402742
网　　址　publish.tju.edu.cn
印　　刷　上海瑞时印刷有限公司
经　　销　全国各地新华书店
开　　本　240mm×320mm
印　　张　6
字　　数　116千
版　　次　2012年6月第1版
印　　次　2012年6月第1次
定　　价　38.00元

MAQ建筑设计事务所——建筑城市化研究
为第四届鹿特丹国际建筑双年展(IABR)设计开发

迪拜破产
棕榈酒店将何去何从?
城市重新规划的宣言

珠美拉棕榈酒店也被称为迪拜棕榈酒店,它不仅仅是最壮观的高档庇护所,在数字、内部空间组织和外部环境关系等方面也是一个建筑典范。这就是为什么该酒店项目既可作为评论的一个研究案例,又可作为一项改造试验。该项目的建造是为了迎合火热的房地产市场;它不是必需品,也不是用来居住的。

当今的奢侈庇护所必将被重新改造。这些庇护所将被划入未来开放城市更大的规划项目中,重新融入周边不断变化的环境中。

为何称这些优质空间为庇护所?从宏观空间角度来讲,很明显这些空间形成被包围的领土,从更大范围的城市或社会中脱离出来。在全球范围内,这些庇护所证明社会向片段式和分了等级的城市化发展的趋势,史蒂芬·格雷厄姆和西蒙·马尔文将这种趋势准确地描述为片段式城市化。

目前所面临的任务是如何将这些庇护所从一个传统自治都市(要塞镇)转变成为一个自治市镇,即城市景观中的一个完整功能区域。

MAQ - architecture urbanism research
conceived for the 4th International Architecture Biennale Rotterdam (IABR)

Dubai is bankrupt
What will happen with the Palm?
A manifesto of urban readjustment

The Palm Jumeirah, also called the Palm Dubai, is not only the most spectacular upscale refuge but also the paradigm in terms of figure, internal organisation and external relations. This is why it serves as both a case study of critique and a test for transformation. It has been built to speculate on a heated real estate market, not on necessity or to live in.

Luxury refuges of today will be inevitably reclaimed. They will be integrated into the larger context of tomorrow's open city and released into the surrounding dynamic environment.

Why call these premium spaces refuges? From a macro-spatial perspective, it is evident that these spaces form enclaves which withdraw from the wider city, or withdraw from society altogether. Across the world, these refuges have been confirming the tendency towards the development of a fragmented and socially stratified urbanism, which was pertinently described as splintering urbanism by Stephen Graham and Simon Marvin.

The task at hand, now: How to turn the refuge from a traditional burgh (a fortified town) into a borough, a quarter that is a functional and thorough part of the urban landscape.

原别墅规划布局 original villa layout

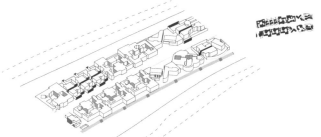

转型后的别墅布局 transformed villa layout

扩建后的别墅布局 extended villa layout

思维景观	THINKSCAPE	02
焦点	ZOOM	06
住宅	RESIDENTIAL	64
设计	DESIGN	80
奖项	AWARDS	90

思维景观
THINKSCAPE

罗马的一天
A day in Rome

Massimiliano and Doriana Fuksas

我们上午九点出发，绕法尔内塞宫的鲜花广场散步，并在Portico d'Ottavia的犹太人糕点店吃了早餐。

我们从一座19世纪的摩尔式建筑Sinagoga前经过，这是罗马一个非常独特的建筑，也是1861年意大利王国创建后建造的第一个新型拓扑结构建筑。

我们回到鲜花广场，然后前往由波罗米尼重建的斯帕达宫(建于1548–1550年)。波罗米尼曾创建了有关封闭庭院内错视伪透视法的杰出作品。我们还访问了珍藏着众多杰出艺术作品的贝尔尼尼艺术馆。

随后，我们前往法尔内塞宫，这里目前已经成为法国驻意大利大使馆。小安东尼奥·达·桑迦洛、米开朗琪罗、雅格布·达·维尼奥拉和雅各柏·德拉·波尔塔等一些16世纪最著名的意大利建筑师曾参与该建筑的建造过程。

16世纪末，波伦亚著名油画家安尼巴莱·卡拉奇完成了法尔内塞画廊重要的连环壁画——《众神之爱》。

我们从法尔内塞宫回到朱利亚大街。这条街道上拥有一些意大利文艺复兴时期最有趣的建筑。

我们在Santa Maria dell'Orazione e Morte教堂前驻足，然后来到法尔科涅里宫，这里目前已经成为Ungheria学院的校区。建筑的部分立面由波罗米尼设计，走廊面对着河流，建筑内部设有楼梯和漂亮的天花板。

圣·卡特里娜教堂大概是巴达萨尔·佩鲁齐的作品，后人于18世纪对其进行了翻新。在左侧我们沿着Sant Eligio大街来到由拉斐尔设计的Sant Eligio degli Orefici小教堂。

教堂的立面设计采用了许多插入元素。

我们回到朱利亚大街，来到由杜梅尼科·丰塔纳于18世纪设计的Spirito Santo dei Napoletani教堂，教堂的建筑立面于19世纪进行了改建。

继续向前走，前面是特里布纳利宫，该建筑由建筑师伯拉孟特开始设计，但从未完成。随后，我们到达Santa Maria del Suffragio教堂(建于1668–1680年)，其米色大理石立面由卡洛·伦纳迪设计。

位于66号的Palazzo Sacchetti可能是由小安东尼奥·达桑迦洛设计完成的。93号是一座16世纪的建筑，采用保罗三世创作的锁眼造型。拉斐尔馆位于85号，这位著名的油画家拉斐尔曾想将其住宅建于此地。

最后，我们参观了由教皇利奥下令修建的San Giovanni dei Fiorentini教堂，该建筑由小安东尼奥·达·桑迦洛、巴达萨尔·佩鲁齐、米开朗琪罗、拉斐尔、雅各布·桑索维诺和波罗米尼合作设计。

在周边地区，我们可以在特莱维喷泉附近的Trattoria Al Moro或Via dei Giubbonari大街的Roscioli吃午餐。

晚餐我们可以在Via dei Chiavari大街上的Ristorante San Lorenzo享用美味的鱼类菜肴。

We start at 9 o´clock and walk around Palazzo Farnese Campo dei Fiori. We have breakfast at the jewish pastry in Portico d'Ottavia.

We pass in front of the Sinagoga, a Moresque 19th century building, which is a quite original architecture in Rome. It is the first architecture to be built after the creation of the Italian State in 1861 and a new topology.

We come back to Campo dei Fiori. We go to Palazzo Spada (built in 1548-1550), which was renewed by Borromini, who created the masterpiece of trompe-l'oeil false perspective in the covered courtyard. We also visit Galleria Bernini which hosts many masterpieces.

Later, we go to Palazzo Farnese which currently houses the French Embassy in Italy. Its building history involved some of the most prominent Italian architects of the 16th century: Antonio da Sangallo the Younger, Michelangelo, Jacopo Barozzi da Vignola and Giacomo della Porta.

At the end of the 16th century, the important fresco cycle of "The Loves of the Gods" of the Farnese Gallery was completed by the Bolognese painter Annibale Carracci.

From Palazzo Farnese we turn to Via Giulia. On this street there are some of the most interesting buildings of the Italian Renaissance.

We stop at church Santa Maria dell'Orazione e Morte. Later, we stop at Palazzo Falconieri, currently the venue of the Academy of Ungheria. The façade was partially done by Borromini .The loggia faces the river and the interior staircase and the beautiful ceilings are well designed.

The church of Santa Caterina, is probably a project of Baldassare Peruzzi and renewed in the 18th century. On the left, if we follow Via di Sant Eligio, we arrive to this little church Sant Eligio degli Orefici designed by Raffaello.

The façade has undertaken numerous interventions.

We are back to Via Giulia and visit church Spirito Santo dei Napoletani designed by Domenico Fontana in the 18th century and its façade was transformed in the 19th century.

If we keep forward we find Palazzo dei Tribunali, which was begun by Bramante but was never finished. We later find Santa Maria del Suffragio (built in 1668-1680) with its travertino façade by Carlo Rainaldi.

At No.66 stands Palazzo Sacchetti, probably designed by Antonio da Sangallo the Younger. At No.93 stands a 16th century house with the escutcheon of Paolo III. At No.85 stands the House of Raffaello built on the site on which the famous painter wanted to build his house.

We can finish with a visit to church San Giovanni dei Fiorentini ordered by Pope Leone decimo with collaborative design by Antonio da Sangallo the Younger, Baldassare Peruzzi, Michelangelo, Raffaello, Jacopo Sansovino and Borromini.

Around the neighborhood, we can have lunch at la Trattoria Al Moro near Fontana di Trevi or at Roscioli in Via dei Giubbonari.
At dinner we can eat good fish at Ristorante San Lorenzo in Via dei Chiavari.

EUR New Congress Center. Building site © Moreno Maggi

EUR New Congress Center. Building site © Moreno Maggi

焦点
ZOOM

轻型透明建筑
Light and transparency

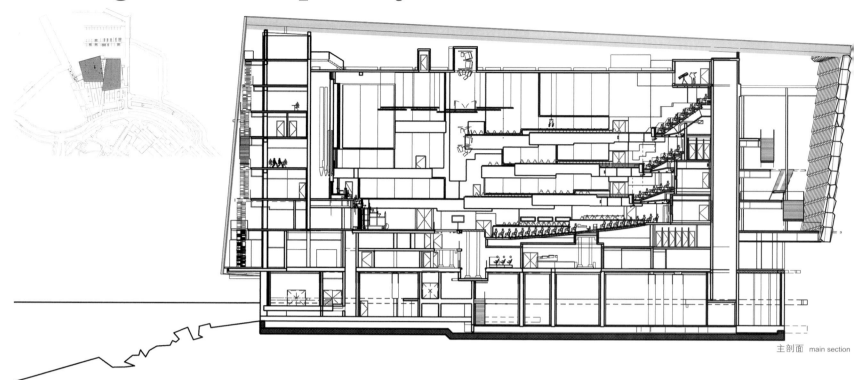
主剖面 main section

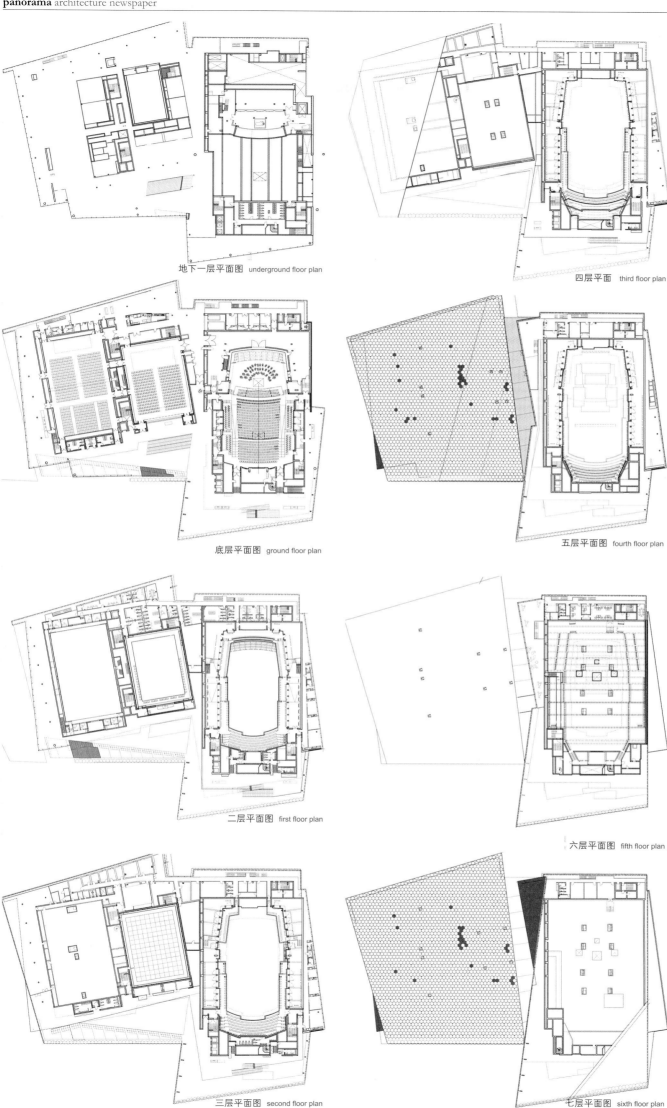

Henning Larsen Architects
"Harpa Icelandic Concert and Conference Hall", Reykjavik, Iceland
"冰岛Harpa音乐厅/会议中心"
雷克雅未克，冰岛

Harpa项目位于雷克雅未克市内一个偏僻而安静的区域，面对广阔的大海和高大的群山。建筑从地面凸起，就像一座大型的光芒四射的雕塑，映衬着蔚蓝的天空和广阔的海港，将该区域转变成一个极具吸引力的城市空间。这里将成为冰岛交响乐团和冰岛歌剧院的新地址。

我们的规划理念是为整个雷克雅未克市打造一个催化剂式的建筑

建筑的玻璃外观真正表现出该结构的独特性。华丽多彩的多面体玻璃立面是著名的建筑师奥拉维尔·埃利亚松和海宁·拉尔森合作创造的完美效果。建筑的南立面采用1 000多块由玻璃和钢材填充的十二边形"类砖石"进行装饰，呈现出缤纷绚丽的彩色效果。其他几个立面和屋顶的部分区域也采用这一几何造型，形成二维平面立面结构。

Situated in a solitary spot, with a clear view of the enormous sea and the mountains surrounding Reykjavik, Harpa stands out like a large, radiant sculpture reflecting both sky and harbor space and transforming the area into an attractive urban space for the citizens. It will become the new home of the Iceland Symphony Orchestra and the Icelandic Opera.

We have aimed at creating a catalyst for the entire city of Reykjavik

It's the building's incredible glass façade that truly defines the structure. The multifaceted glass façades are the result of the collaboration between renowned architects Olafur Eliasson and Henning Larsen. Made of a twelve-sided space-filler of glass and steel called the "quasi brick", the building appears as a kaleidoscopic play of colors, reflected in the more than 1,000 quasi bricks composing the southern façade. The remaining façades and the roof are made of sectional representations of this geometric system, resulting in two-dimensional flat façades.

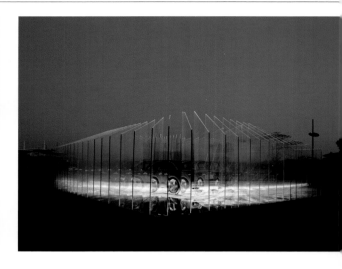

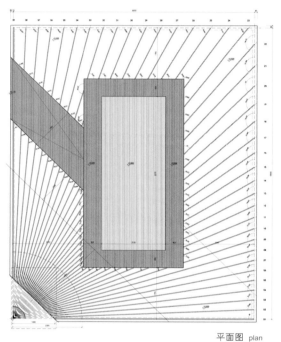

平面图 plan

动态展馆
A dynamic pavilion

非常建筑工作室 Atelier FCJZ
"Audi Haus", Shanghai, China
"奥迪展场",上海,中国

中国建筑师们打造了一个由玻璃隔板组成的结构,里面是一辆被玻璃板所包围的奥迪轿车。当参观者围绕展品移动时,会感受到动态旋转的幻觉效果。该展场建筑正在中国上海国际F1赛车场进行展示。

Chinese architects have created an installation consisting of glass partitions encompassing an Audi vehicle inside, creating the illusion of movement and rotation as the viewer walks around the display. It has been on display at the F1 international racing circuit in Shanghai, China.

通过对玻璃板的布置可对轿车产生多重反射效果

参观者可以走进展馆近距离观赏这款车型,并亲身体验展厅内部梦幻般的效果,其中的玻璃窗格将外界环境变得模糊。

The glass panels are arranged to produce multiple reflections of the car

Visitors could walk inside the pavilion to have a closer view of the car and experience a more intimate setting where the glass panes blur the outside surroundings.

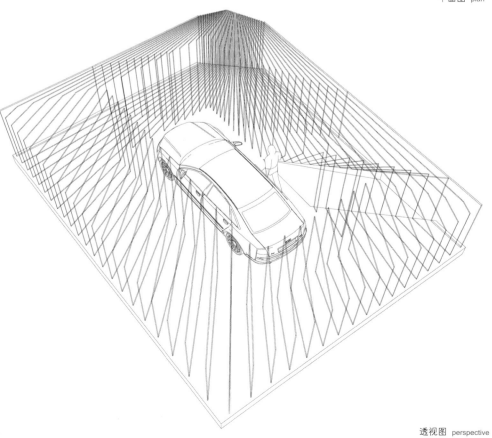

透视图 perspective

从汽车到公交车辆
From car to public transport

KCAP Architects&Planners
"P+R facilities De Uithof", Utrecht, Netherlands
"P+R De Uithof停车场"，乌德勒支，荷兰

P+R De Uithof停车场项目已经开始施工。项目设计由KCAP建筑师和设计师事务所与SK工作室和乌德勒支Ingenieursbureau建筑事务所合作完成，它能够提供2 000个停车位，将成为乌德勒支最大的停车场。本项目计划于2013年竣工。

Construction work has started for the P+R facilities De Uithof. The design developed by KCAP Architects&Planners in cooperation with Studio SK and Ingenieursbureau Utrecht will be the largest parking of Utrecht with 2,000 parking spaces. Completion is planned in 2013.

该项目的特别之处在于其在地面上将公共汽车/有轨电车车站与建筑整合成一个整体

What makes the project special is the integration of a bus/tram station into the building on ground level

该建筑共10层，为医院、大学、应用科技大学及大众提供停车空间。项目还包括一个自行车棚，并作为一个过渡空间将快速公交系统与市中心连接起来。

The 10-storey building offers parking facilities for the hospital, the university, the Hogeschool and public use. It contains a bike shelter and serves as a transfer connecting the fast public transport to the city center.

减少碳足迹
Mitigating carbon footprint

Michael Jantzen
"The Black Hole Research Center"
"黑洞研究中心"

该项目是对位于干燥炎热地区的一个大型太阳能建筑的概念设计。结构外观设计类似一个位于大型螺旋状星云中央的黑洞。

It is a conceptual design proposal for a large, solar powered building designed to be located in a hot dry area. The structure's design is symbolically based on the image of a large black hole located in the center of a large spiral galaxy.

The symbolic black hole in the center of the structure is actually a very large array of photovoltaic solar cells that would power most of the research center. In clear allusion to a black hole pulling light, the grandiose solar array of the center will absorb light, and then convert it into a final form of usable and sustained harnessed energy.

轻型生态混凝土复合材料

结构中央的黑洞实际上是一组巨大的光伏太阳能电池阵列，将为研究中心的大部分建筑提供电能。与黑洞吸引光线的性质相比，这个位于中央的巨大的太阳能电池阵列将吸收太阳光，并将其转化为一种可持续使用的能源。

Light, eco-friendly, concrete composite materials

© Studio SK

© Studio SK

超现实主义空间
A surreal space

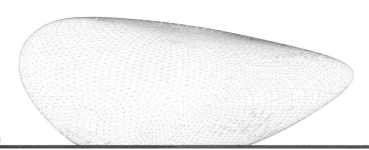

立面图 1 elevation 1

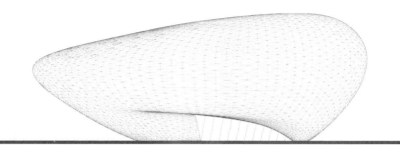

立面图 2 elevation 2

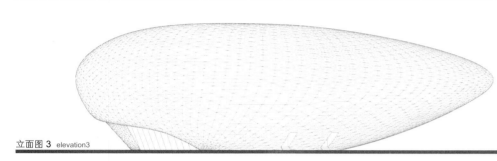

立面图 3 elevation3

立面图 4 elevation 4

朱锫建筑事务所 Studio Pei-Zhu
"OCT Design Museum", Shenzhen, China
"OCT设计博物馆"，深圳，中国

该项目的设计灵感来源于临近海湾的地理环境以及该项目的功能要求，即提供一个可以展示设计作品的超现实主义空间。闪亮的银色外观上分散着三角形的小窗户，看起来像是鸟儿正在云层或浓雾中飞翔。

The inspiration for the project comes from both the location close to the bay and the needs of the program, a surreal space for design exhibitions. The silver shiny exterior with scattered triangular small windows seems to be part of a cloud or dense fog with birds in flight.

外观像一个飞碟

Reminiscent of a flying saucer

在一般情况下，一辆汽车看起来很沉重，但在如此广阔的空间里，汽车则变得非常渺小。该建筑的轮廓曲线、阴影以及强烈的颜色成为人们视线中的焦点。

Typically an automobile looks very heavy but in this limitless space it becomes weightless, letting its curves, shadows and intense colours become the focal point of the show.

屋顶层平面图 roof plan

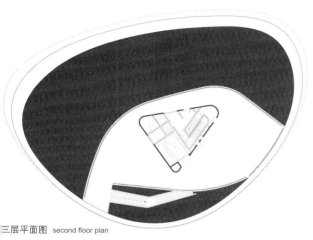
三层平面图 second floor plan

二层平面图 first floor plan

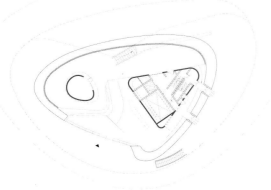
底层平面图 ground floor plan

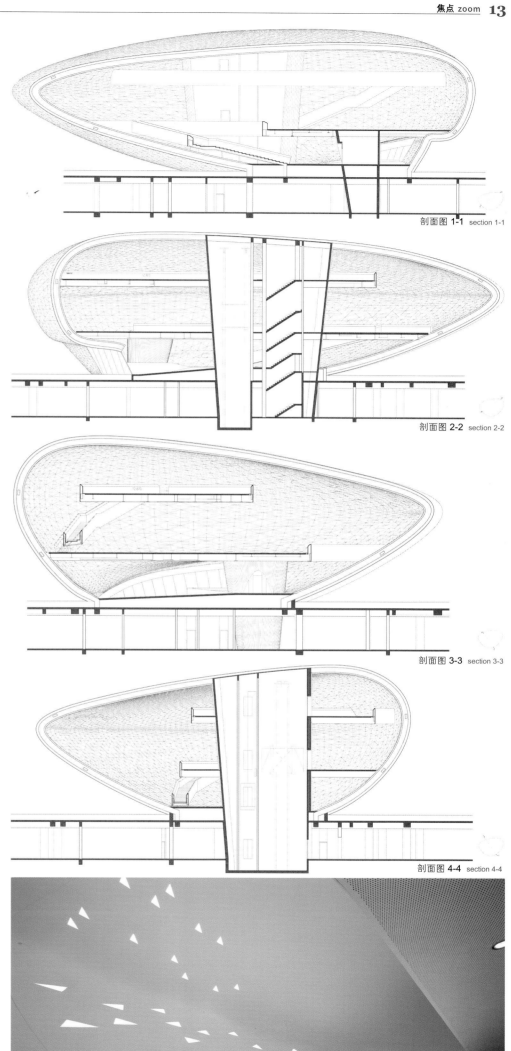

剖面图 1-1 section 1-1

剖面图 2-2 section 2-2

剖面图 3-3 section 3-3

剖面图 4-4 section 4-4

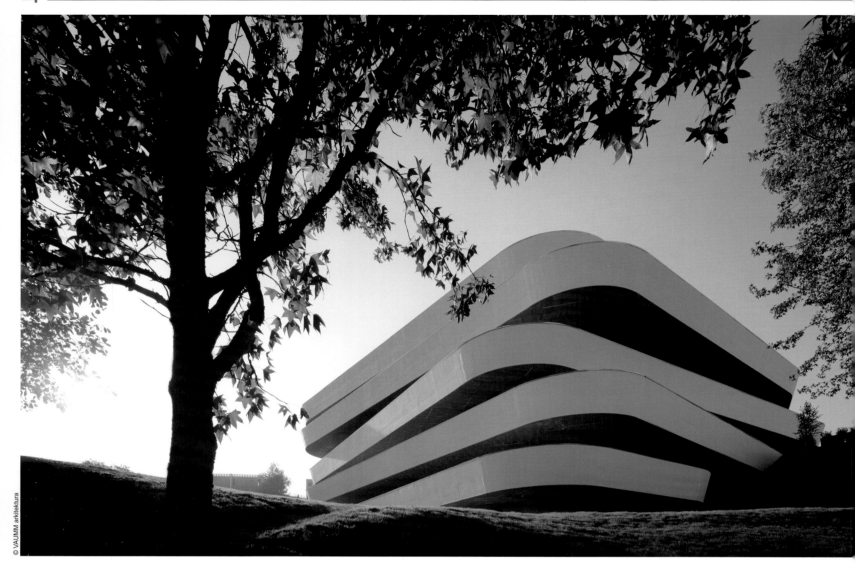

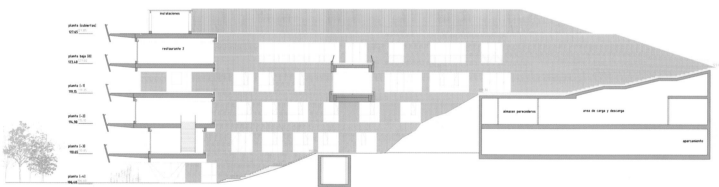

剖面图 section

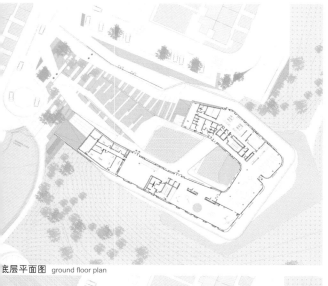

底层平面图 ground floor plan

地下一层平面图 floor plan -1

地下二层平面图 floor plan -2

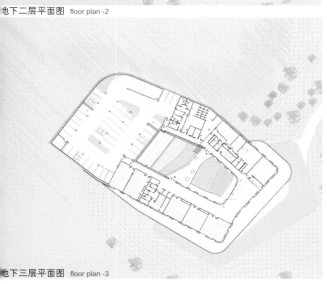

地下三层平面图 floor plan -3

地下四层平面图 floor plan -4

首先用视觉享受佳肴!
You eat with your eyes first!

VAUMM arkitektura
"Basque Culinary Center", Donostia-San Sebastián, Spain
"巴斯克烹饪中心",圣塞巴斯蒂安,西班牙

该建筑是美食科技大学和烹饪研究创新中心位于西班牙圣塞巴斯蒂安的总部。这所新烹饪学校曾经希望有一座巨大而先进的建筑为学生们提供宽敞的厨房、工作室、初学者实验室、报告厅、娱乐室和自助餐厅等设施。

The building is the head office for the Gastronomic Science University and Research Innovation Center for Gastronomic Science in Donostia-San Sebastián, Spain. The new cooking school wanted a huge, state of the art facility to give its students large kitchens, workrooms, labs for hands on learning, theaters for lectures, lounges and cafeterias.

可持续性技术与结构设计
Technically and structurally sustainable

天津西站位于北京市西南方向大约130千米处,是北京至上海高速铁路中的一个重要站点。该项目共设24个站台,使地区线路与高速路网相互连接。

它如今是一个重要站点

建筑中央覆盖着一个巨大的钢架拱形穹顶,穹顶采用钻石状玻璃进行装饰。拱形屋顶使人们联想到一扇巨大的城门。

The new intersection, which is located about 130 kilometers south-west of Beijing, serves as an important station along the high speed rail line between Shanghai and Beijing. The 24-platform station connects regional lines to the high speed network.

It now serves as an important station

The concourse is a striking space covered by a vaulted ceiling of diamond-shaped steel and glass roof construction. Its curved roof is reminiscent of a large-scale city gate.

gmp Architekten
"West Station", Tianjin, China
"天津西站",天津,中国

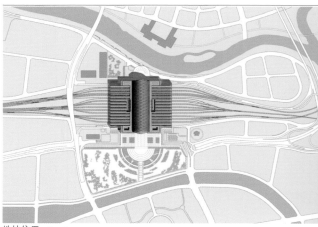

地块位置 site plan

Garmendia Arquitectos
"Sports Centre Barakaldo", Cruces, Bizkaia, Spain
"巴拉卡尔多体育中心"，克鲁塞斯，比斯开，西班牙

简洁大方，经济实用
Simple and economical

该项目包括一个学校的运动场，项目用地是一个退台式地形。每层退台在不同高度上都设计了不同的运动场地。

The project covers a playing field within a school environment characterized by the existence of a terraced site. The site's stepped terraces feature different playgrounds at different heights.

涂以鲜亮颜色的标志性顶盖

Iconic and brightly coloured canopy

设计师与业主决定打造一座类似于"帽子"形状的建筑，为巴斯克地区的孩子们提供玩耍和避雨的空间。

Designers and customers decide to create a "hat"-shaped building to give children a place to play by covering them from the usual rain of Basque Country.

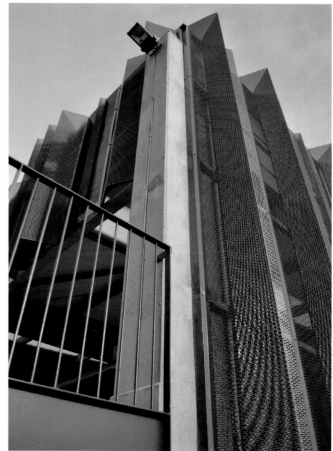

Hikaru Kinoshita＋TOFU＋Yuji Tamai
"Osaka Board Game Park",Osaka,Japan
"大阪棋盘游戏公园"，大阪，日本

该项目的目标是以非常有限的预算为大阪居民建造一个临时娱乐空间。但遗憾的是，该空间只是临时的，将来会被拆除，在这里再建造其他项目。

The aim of this project is to make a temporary recreation place for residents of Osaka at a very modest budget. Unfortunately, it is temporary and will be dismantled, and in its place there will be something else.

公共互动空间

An interactive public space

不仅仅是实用的结构
More than utilitarian structures

Brooks + Scarpa
"Santa Monica Municipal Parking Garage", Los Angeles, USA
"圣莫尼卡市政停车场"，洛杉矶，美国

Brooks+Scarpa建筑师事务所受命对圣莫尼卡市内第三步行街周围的八个停车场进行优化设计。其中第7、8两个停车场临近由著名建筑师弗兰克·盖里(Frank Gehry)在20世纪80年代设计的圣莫尼卡购物中心，并为该中心提供停车位。该中心后来被Jerde合作伙伴公司领导的设计团队改造为一个户外步行购物街。

更加具有步行街的气氛

停车场实际上就是一些巨大的显示墙，这是设计中开发的一个新特色，完美展示了鲍尔-诺格斯和安妮-玛丽·卡尔森等公共艺术家们的优秀作品。

Brooks + Scarpa was commissioned to design improvements to eight parking garages surrounding the popular 3rd Street Promenade of Santa Monica. The parking garages 7 and 8 in particular, adjoin and serve the Frank Gehry-designed Santa Monica Place, a 1980s indoor shopping mall that was transformed by the Jerde Partnership-led team into an open-air shopping promenade.

A more walkable and street-like atmosphere

They are essentially giant display walls, and the new design exploits this feature, showcasing the outstanding work of accomplished public artists such as Ball-Nogues Studio and Anne-Marie Karlsen.

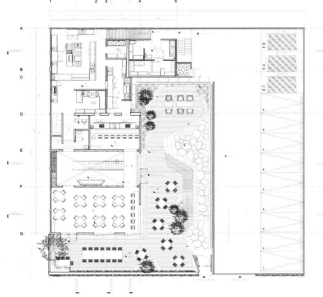

过滤灯光
Filtering light

rojkind arquitectos-Esrawe Studio
"Tori Tori Restaurant",
Mexico City, Mexico
"Tori Tori餐厅",墨西哥城,
墨西哥

Tori-Tori餐厅被认为是墨西哥城最好的日本餐厅之一,目前已经搬到墨西哥城波兰科地区更大的场所。一旦进入该餐厅,你会发现自己在一个露台上,这里的食物和饮品都被自然植物所包围。建筑立面和景观被细心设计成餐厅的一个延伸部分。

动态流线型立面

建筑周围设计了整套的桌椅来增强每个内部区域的用餐气氛。

Considered as one of the best Japanese restaurants in Mexico City, Tori-Tori has been moved to a bigger location in Polanco, Mexico City. Once you enter it you'll find yourself in a terrace, where eating and drinking are embraced by natural vegetation. The building's organic façade and landscape were carefully designed to become an extension of the restaurant.

Façade appears to waver and curve

A complete collection of tables and chairs were created to enhance dining atmosphere in each interior area.

底层平面图 ground floor plan   二层平面图 first floor plan   三层平面图 second floor plan

锐利的边缘
Sharp edges

Alcmea Architects
"Wood Mass", Aubergenville, France
"木块建筑"，
欧贝尔让维尔，法国

This 500 m² wooden building is to house the new headquarters and studio of a young electricity firm. Referring to the classical formalism of industrial buildings, particularly present in the area where the project is located, "Wood Mass" seeks to distinguish itself by the treatment of details and clean lines.

这座占地500平方米的木质建筑将成为一个年轻的电力公司的新总部和工作室。参照工业建筑的古典形式主义并考虑当地环境，"木块建筑"试图通过细节设计和干净的线条设计使自己与众不同。

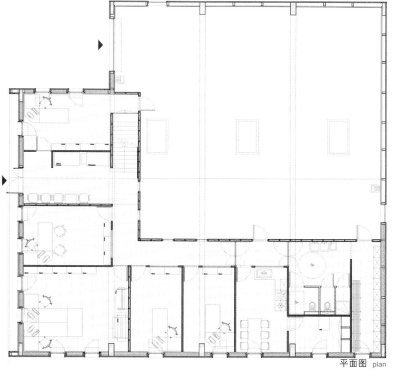

平面图 plan

结构功能和表面处理的新机遇使木材重新成为主要的建筑材料

New opportunities for structural use and surface treatment are restoring wood to the prominent and visible role

Inside, workshop spaces and offices communicate with each other, physically and visually to create a spatial homogeneity and promote exchanges among teams.

建筑内部的车间和办公室在功能和视觉上相互呼应，创造均质性空间结构，促进团队之间的交流协作。

Rafael de La-Hoz
"A.M.A. Insurance Headquarters", Madrid, Spain
"A.M.A.保险公司总部",
马德里，西班牙

根据总体规划的要求，建筑物的底层要尽可能小地占用地面空间，以保留商务公园景观。因此，该项目的设计理念是将建筑的侧立面抬起，从而创造一个升离地面的大型建筑结构。

金属立面为该区域带来未来主义元素

该建筑采用两个25平方米的巨大悬臂，使景观可以沿建筑向下蔓延。建筑立面清晰展示设计结构体系。沿对角线布置伸张器，从而显示出玻璃立面框架结构。

Determined by the master plan, the building takes up the ground space as little as possible to preserve the Business Park landscape. This building is conceived as a volume that raises its side façades to creat a heavy mass elevated over the ground.

A metallic façade that brings futurism to the area

Two huge 25㎡ cantilevers were used to allow the landscape to spread down the building. The façades clearly show the structural system of its design. In this way, a pair of tensors are placed diagonally to show the skeleton of the glass façade.

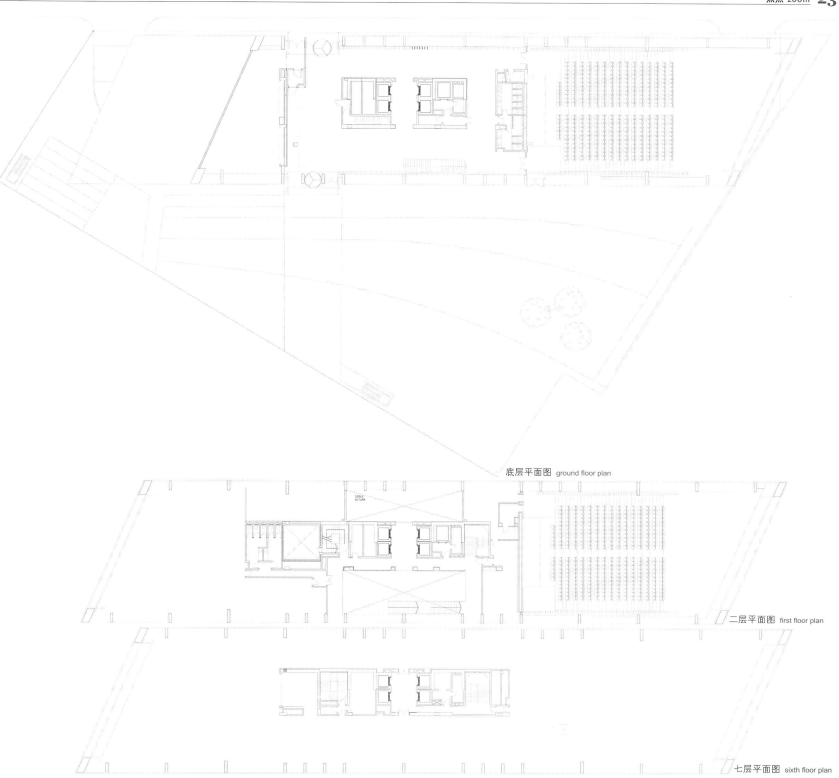

底层平面图 ground floor plan

二层平面图 first floor plan

七层平面图 sixth floor plan

商店与酒店一体化
It integrates store with hotel

Goettsch Partners
"Mixed-use Complex",
Chengdu, China
"多功能建筑综合体"，
成都，中国

该建筑综合体临近成都市中心商务区内一个充满活力的公共广场，占据一整块城市街区，包括一个12层的百货商店，上面是25层的君悦大酒店。该酒店成不对称形状堆叠在零售商店裙楼上面，使该建筑体量与周围建筑一体化。

中国经济指数式的增长也保证了对此类多功能项目的需求

酒店塔楼坐北朝南，为所有房间提供无障碍视野和丰富的自然光。酒店里的百货商店是中国中部地区最大的百货商店之一，该酒店扩建后将能够提供450套房间。该酒店预计于2012年底试营业。

Occupying a full city block adjacent to a vibrant public square in the central business district of Chengdu, the complex is anchored by the 12-story department store, with a 25-story Grand Hyatt Hotel above. By asymmetrically stacking the hotel tower above the retail podium, the design helps integrate the building's mass with its surroundings.

China's exponential growth also guarantees the demand for these types of multi-use projects

The hotel tower, sited to the south, provides all rooms with unobstructed views and abundant natural light. The department store is one of the largest stores in central China, and the hotel will provide 450 rooms at full build-out. A scheduled soft opening for the hotel is planned for late 2012.

群光广场段剖面图 chicony plaza section

Nabito Architects
"Sensational Garden",
Frosinone, Italy
"感官花园"，
弗洛西诺，意大利

意大利弗洛西诺市考尔索·拉齐奥的邻居们终于能够享受他们期待了25年之久的第一个公共空间了。这个感官花园是一个整体规划方案，对公共空间和服务设施进行更新，并将它们统一到周围的住宅环境中。该项目的目标是为用户提供一个景色多样化的步道环境。每个区域分别代表人类的五种感官功能之一。

居民终于能够享受他们的第一个公共空间了

自然元素(树木、灌木和鲜花)与人工元素(水泥和树脂)之间的均匀搭配不仅使该花园易于维护，同时可以提供持久多样化的景观。

人工元素与自然元素的融合
Between artificial and natural elements

The neighbourhood of Corso Lazio, in the city of Frosinone, Italy, can finally enjoy its first public space that it has been expecting for 25 years now. The sensational garden represents the starting point of a big masterplan for the renewal and integration of the public spaces and the services into the housing neighbourhood. The goal of the project is to invite users to a path in which the scene is always changing. Each area is a metaphor of one of the five human senses.

Citizens can finally enjoy their first public space

The balanced blend of the natural essences (trees, shrubs and flowers) and the artificial elements (cement and resin) makes the garden not only easy to maintain but also simultaneously durable and mutable.

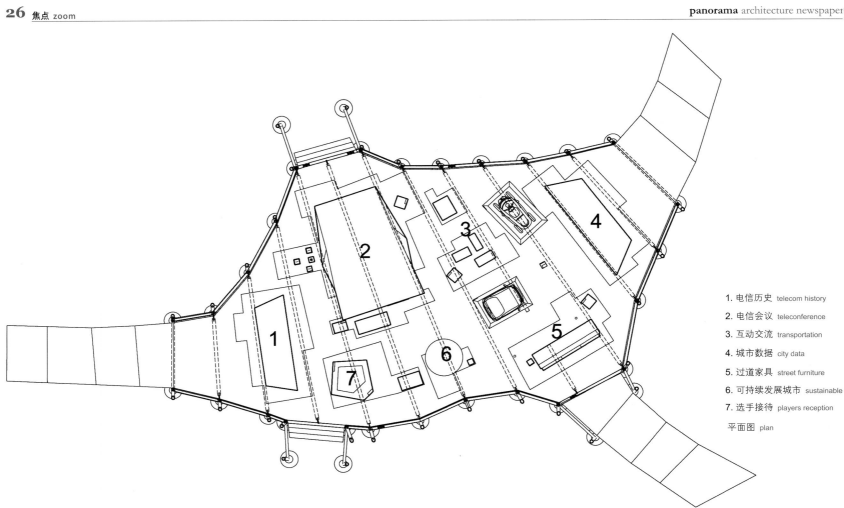

1. 电信历史 telecom history
2. 电信会议 teleconference
3. 互动交流 transportation
4. 城市数据 city data
5. 过道家具 street furniture
6. 可持续发展城市 sustainable
7. 选手接待 players reception

平面图 plan

一项挑衅式的设计理念
A provoking concept

Jakob + MacFarlane
"La Ville Intelligente",
Parc de la Villette, Paris, France
"智能城市",
德拉维莱特公园,巴黎,法国

本项目采用由先张法法拉利帆布和管状钢棒构成的开放式结构,钢棒之间巧妙地连接形成一个形状优美的外壳。该"智能城市"项目原来是为Helldemain博览会开发设计的,目前临时在巴黎德拉维莱特公园里进行展示。

This open structure is an ensemble of pre-tensioned Ferrari canvas and tubular steel rods that have been cleverly joined together to create an exquisite outer shell. La Ville Intelligente was originally developed for the Helldemain exposition and is at present temporarily on display inside the Parc de la Villette of Paris.

简单外形下的未来主义设计手法

Simple in outlook but futuristic in approach

金属方格框架下面紧紧贴附着相同的鲜橙色帆布,其中包含着许多展馆空间。

The interior has the same bright orange colored canvas that is tightly attached beneath the metallic grid, which contains a number of pavilions.

Reiulf Ramstad Architects
"Fagerborg Kindergarten",
Oslo, Norway
"Fagerborg幼儿园",
奥斯陆，挪威

该地区的建筑以19世纪前50年的建筑形式为特点。与周围砖结构的历史建筑形成鲜明对比，该幼儿园建筑统一采用落叶松木芯材的木质外观。该建筑共有四个相互独立的单元，为四至六岁的孩子提供活动空间，该设计方案使四个单元可以根据需要进行分隔与组合。

内外空间相互呼应，实现环境的延续性

整个建筑根据孩子们的身高共设有四种不同的开窗形式，在不同房间营造出多样化的采光效果。

The area is characterized by architecture from the first half of the 19th century. In contrast to the surrounding historic brick buildings, the kindergarten is uniformly dressed with a wooden cladding of larch heart wood. Housing four separate units for children between the ages of one and six, the solution enables the units to function both independently and together as required.

Interior and exterior space work in dialogue to provide a continuous environment

There are four types of window openings appropriate to children's dimensions. This creates a variety of light modulation in the different rooms.

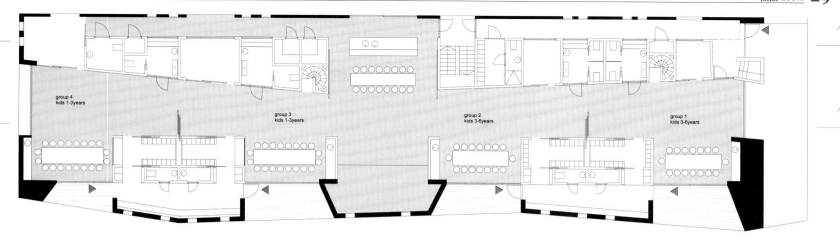

平面图 1 floor plan 1

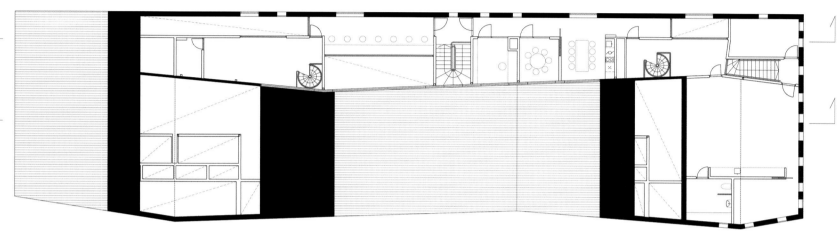

平面图 2 floor plan 2

当代建筑形式
A contemporary form

Allmann Sattler Wappner
"Stachus Passage", Munich, Germany
"施塔舒斯地下通道",
慕尼黑,德国

Stachus地下建筑位于慕尼黑火车站和市中心之间的卡尔斯广场下方,该广场常被人们称为"施塔舒斯"(Stachus),是慕尼黑最重要的交通枢纽之一。除了可以通往地铁站、城铁站和有轨电车车站外,这里还是一个极为繁华的购物中心和就餐区域。

圆形几何图案是天花板的主要特征

利用"交通圆"的环境特点,每天16万人次的客流可以逛逛商店,穿过商店之后在同一内部空间可以方便地到达路面公交、地铁和快速铁路等交通系统。圆形元素设计延续至任何一个细节。

The Stachus underground structure situates at Karlsplatz, better known as Stachus, between the main station and the city centre. It is one of the most important public transportation hubs in Munich. Apart from providing access to the U-Bahn station, S-Bahn station and trams station, the space is an extremely busy shopping mall and food court.

The geometric shape of the circle is the characteristic feature of the ceilings

By making use of the "traffic circle" idea, the daily 160,000 pedestrians can walk round the shop areas, cross them, walk through them and still find their way to the transportation systems of streetcars, underground and rapid-transit railway. The circle segments continue down to every detail.

每天16万客流量
160,000 pedestrians daily

深入大海内部
Inside the sea

鸟瞰图 aerial view

JDS Julien De Smedt
"Harbour Bath and Blue Base",
Faaborg, Denmark
"港口浴池与蓝色基地",
福堡,丹麦

该项目是最近一次设计竞赛的获奖作品,由KLAR事务所、Creole建筑师事务所和Sloth-Moller工程师事务所合作设计完成。浴池的设计理念来源于打造一个公海浴池区域的想法,墩座向海面伸展,围出多个游泳区域。这样就创造了一个"手指状平面图",四个不同宽度和长度的浴池桥梁之间形成多个开放式水池,通过每个桥梁都可以进入水中,而每个水池都具有特定的用途。

It is a winning entry of a recent competition designed in collaboration with KLAR, Creole Architects and Sloth-Møller engineers. The concept of the bath is based on the idea of creating an open sea bathing area with piers branching out seawards to create swimming areas between them. This creates a "finger plan" with open basins between four bath bridges of different width and length, where each bridge offers a new way to get into the water in which each pool has a specific use.

吸引当地居民前来戏水和参加水上运动

It will invite the locals to swim and enjoy water sports

木质桥墩形成坡道、楼梯、休息区和小型儿童水池。港口浴池设有多个更衣室、一个桑拿浴室以及一个300平方米的陆上乘船旅游集合地。

The wooden piers form ramps, stairs, sitting area and small pools for children. The Harbour Bath has changing rooms, a sauna and a 300 m² meeting point for boat tourism on land.

 POOLS
 HANDICAP ACCESSIBILITY

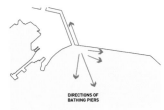 DIRECTIONS OF BATHING PIERS
 PAVILIONS
 AMBULANCE PASSAGE

 PROGRAMMATIC DISTRIBUTION
 GENERAL ACCESS
 PAVILION ENTRANCE

平面图 plan

都市实践 URBANUS
"Maillen Hotel & Apartment", Shenzhen, China
"Maillen酒店式公寓",深圳,中国

该项目的空间布局使人们想起中国传统式村庄和山峦风景。自然与建筑之间的关系变得不再那么清晰,屋顶采用斜坡式设计,居住者进入内部空间会发现自己身处植物庭院中。

Spaces and form evoke the imagery of traditional Chinese villages and mountainous landscapes. The relationship between nature and construction is blurred as dwellers enter vegetated courtyards inside. The roof is sloped.

无缝过渡
Seamless transition

区块位置 site plan

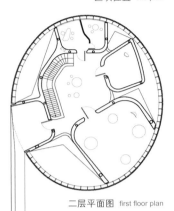

二层平面图 first floor plan

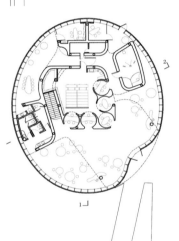

底层平面图 ground floor plan

新标识
A new identity

都市实践 URBANUS
"Nanshan Wedding Center",
Shenzhen, China
"南山婚姻登记处",
深圳，中国

该项目是中国民政大厅的新建筑，拥有独特的外观造型。

The new structure for China's Department of Civil Affairs boasts a unique aesthetic.

试图唤起人们结婚登记时的兴奋心情

An attempt to return excitement to the legalities of registering as a married couple

"南山婚姻登记处项目是一种新建筑类型。"都市实践建筑事务所项目负责人张震解释说，"一方面，我们希望给结婚登记过程带来新的标识性建筑；另一方面，为城市其他用户提供不同类型的公共空间。"

"The Nanshan Wedding Center is a new architectural typology," says project director Zhang Zhen of URBANUS Architecture & Design. "On one hand, we hope to bring a new identity to the process of marriage registration. On the other hand, it provides different types of public space for other users from the city."

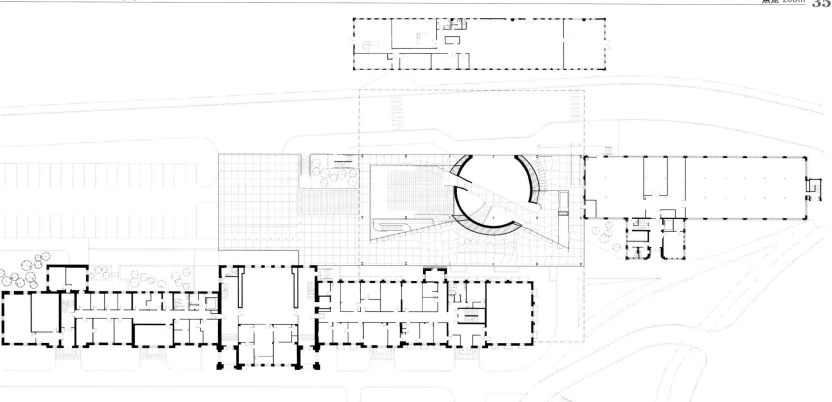

平面图 plan

OMA
"Milstein Hall", Ithaca,
New York, USA
"米尔斯坦会堂", 伊萨卡,
纽约, 美国

康奈尔大学建筑艺术与规划学院 (APP) 米尔斯坦会堂占地 4 366.4 平方米。这个盒状建筑将一系列建筑统一成一个连续的空间。新设施既注重项目之间的相互作用，同时也注重将来规划的灵活性。

The new Milstein Hall provides 4,366.4 m² for Cornell's College of Architecture, Art and Planning (APP). The box transforms a cluster of buildings into a continuous space. The new facility focuses on stimulating the interaction of programs while allowing flexibility over time.

升起的盒子
A lifted box

作为一种连接元素

It acts as a connector

在水平方向对截面进行巧妙处理，以此划分不同的功能区域。整个建筑采用悬空设计，从学院校园的地面升起。

Within the horizontal plate, subtle manipulations of the section define zones that suggest and trigger particular uses. Lifted, the building leaves the figure-ground of the AAP campus.

36 焦点 zoom

引人瞩目
Undoubtedly visible

DOMINIQUE PERRAULT ARCHITECTURE
"Citylights",
Boulogne-Billancourt, France
"城市之光",
布洛尼-比扬古,法国

"城市之光"项目是Pont de Sèvres老建筑进行完全重新都市化规划过程的开始,包括三个建筑,扩建改造以后可以提供八万平方米的办公空间。

该项目面临的最大挑战是将19世纪70年代的建筑改造成为21世纪的地标性建筑

"我们将该项目命名为'城市之光',是因为该建筑将成为该区域内一个视觉标志或一个明亮的灯塔,从远处布洛尼-比扬古市入口处就可以看见该建筑。"

Citylights starts a process of complete re-urbanization from the elder Pont de Sèvres, consisting of three buildings totaling over 80,000m² of office space inside a major restructuring and expansion.

This project is a real challenge: to transform this set from the 1970s into a 21st century landmark building

"We chose this name because 'Citylights' will be a visual cue or a bright beacon, visible from afar at the entrance of Boulogne-Billancourt."

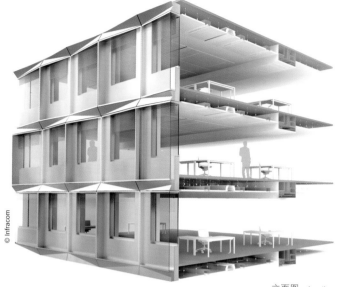

立面图 elevation

立面图 elevation

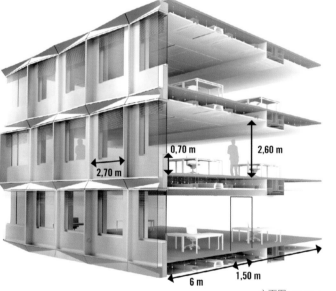

立面图 elevation

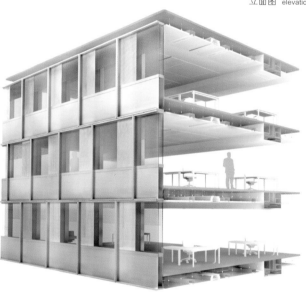

立面图 elevation

环保建筑
Environmentally friendly

Michael Jantzen
"Solar Winds Cultural Arts Center"
"太阳能风能文化艺术中心"

该项目的设计理念是打造一个利用太阳能和风能的大型建筑，为多样化的文化艺术活动提供空间。

The center is a design proposal for a large solar and wind powered structure dedicated to a wide variety of cultural arts activities.

展示如何实现替代能源与大型公共建筑一体化

It will help to demonstrate how to integrate alternative energy into large public buildings

该文化艺术中心由七个圆锥形模块化建筑组成，它们通过基座相互连接到一起。每个建筑都配备了大型竖轴式风力涡轮机，与太阳能电池和太阳能热量提取系统一起提供文化艺术中心所需的所有电能以及供暖和热水供应所需的能量。

The center is composed of seven conical shaped modular structures that are merged together at their bases. Each of the seven is fitted with a large vertical axis wind turbine. The wind turbines, solar cells and solar heat extraction systems provide all of the electricity, space heating and water heating energy needed to operate the center.

Schmidt Hammer Lassen Architects
"Campus Aarhus N,
VIA University College", Denmark
"VIA 大学 Aarhus N 学院校园"
丹麦

该建筑将以前位于Aarhus学院不同区域的VIA学校卫生教育部门集中到一起。校园建筑由四个翼楼组成，将中间围成的一个具有良好灯光照明和动态线条的中庭作为中心广场。建筑的核心功能布置在广场的四周，其中包括礼堂、多媒体中心、食堂以及其他共享设施等。

The building unifies VIA's healthcare educations previously located at different places in Aarhus. The campus building consists of four wings which centre on a light filled and dynamic atrium–a central square. The core functions are placed around the square–auditoriums, a multimedia centre, the canteen, and other shared facilities.

中心广场是该建筑的空间标识

The central square is the spatial identity of the building

挺拔的外墙采用锈红色低合金高强度钢框架，与立面上特意设计的常青藤绿化形成鲜明对比。

In the exterior the robustness is shown in the strong, rusty-red corten steel frames which form a beautiful contrast to the ever green band of ivy in the façade elements, designed especially for this building.

崇尚教育
Inspiring for education

立面图 elevation

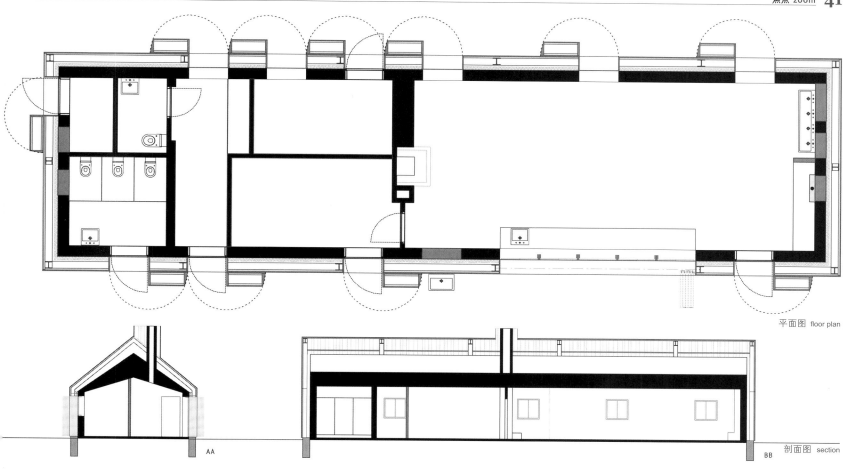

平面图 floor plan

剖面图 section

独特的建筑
A unique installation

MLRP
"Mirror House",
Copenhagen, Denmark
"镜面房屋",
哥本哈根,丹麦

"镜面房屋"为幼儿园提供灵活的教室空间和休息室。建筑外墙用带焦色的木材包裹,端部采用抛光钢板进行装饰,反射出周围的运动场和树木。MLRP建筑事务所将公共公园内一个被涂鸦的废旧建筑成功地改造成一个完全互动式的建筑,同时与周围的自然环境无缝连接。

The Mirror House is a flexible space and restroom used by kindergarten classes. The pavilion is claded by charred timber but its polished steel ends reflect the surrounding playground and trees. MLRP has managed to turn an old, dilapidated building located in a public park, which was decorated with by graffiti, into a fully interactive building, which at the same time seamlessly connects with the adjacent natural area.

像是传统游乐场中的娱乐设施

It is like a traditional attraction at funfairs

夜晚,关闭百叶窗后,建筑就会隐藏在一片漆黑当中。

At night the shutters are closed, making the building anonymous.

现代化儿童中心
Modern children center

Crossboundaries Architects
"Family Box", Beijing, China
"家盒子"，北京，中国

该项目的设计出发点是为成人和儿童设计不同尺寸和高度的空间。但是如何在两者之间寻找平衡呢？当Crossboundaries Architects接管该项目时，项目正在建设中，由于项目早期有另外一家设计公司参与，所以项目采用了不同的设计功能，即现有结构系统、柱网和建筑基础。各种各样的活动都被"装入"这些独立的盒子当中。

运动场与幼儿园的统一体

Designing spaces of different sizes and heights for adults and children is the starting point of this building. How to find an inspiring balance between them? When Crossboundaries Architects took over the project, the building was already under construction, designed for a different function an existing structural system, column grid

It is something between a playground and a kindergarten

and the building footprint—due to the involvement of another design firm in an earlier project stage. The various activities are enclosed in freestanding boxes.

二层平面图 first floor plan

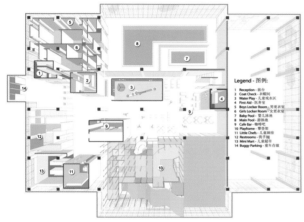

底层平面图 ground floor plan

地下一层平面图 basement floor plan

波纹绸效果
Watered-silk effect

Tom Price
"Tower and Wall", Gloucester, UK
"高塔与墙壁"，格洛斯特，英国

Kimbrose广场是格洛斯特市中心新改建的一个公共空间。根据Ginkgo Projects的规划设计，汤姆·普莱斯(Tom Price)创造了两个永久性公共空间来反映该区域的悠久历史，其中包括一个16米高的高塔和一个30米长的墙壁。

安静的景观与功能空间

从远处观看，这个钢制高塔仿佛在上下浮动。高塔的外观造型仿照起伏的水波纹进行设计，像是从上倾泻下来的水柱，象征Kyneburgh喷泉下落到井中。高塔的内表面采用螺旋上升设计，像是一团上升的蒸汽，使人们联想起"时光隧道"。

Kimbrose Square is a recently transformed area of public space in Gloucester's city centre. Curated by Ginkgo Projects, Tom Price created two permanent public artworks to reflect the area's fascinating history, comprising a 16-metre high tower and a 30-metre long wall.

Both a spectacle and a place for quiet contemplation

When approached from a distance, the tower, though made of steel, appears to move up and down. The outer shape is based on an undulating body of water, as if poured from above, and represents the fall of Kyneburgh's body down the well. The tower's inner surface spirals upwards, like a rising body of steam, and makes people recall the "tunnel of time".

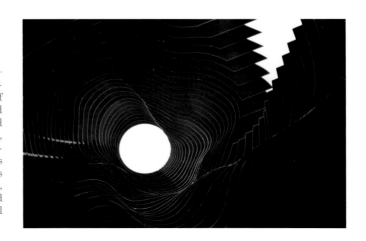

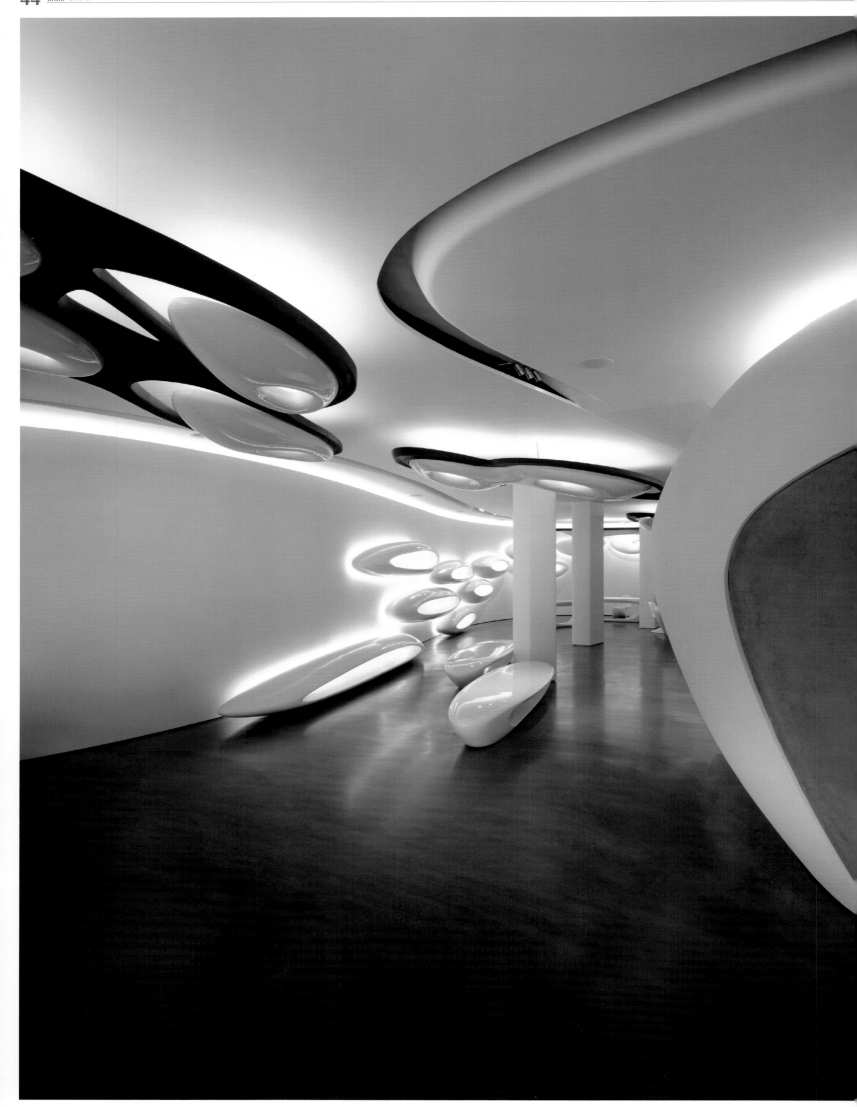

不仅仅是一个展示空间
Much more than just a display space

Zaha Hadid Architects
'Roca London Gallery',
London, UK
"乐家公司伦敦展廊"，伦敦，英国

乐家公司伦敦展廊位于切尔西港口附近的皇家码头(Imperial Wharf)，它不仅仅是一个纯粹的视觉景观，还利用精度和控制手法帮助参观者了解空间建筑与乐家浴室用品设计之间的关系。

在伦敦的第三个项目

从流水形态中获得灵感，乐家公司伦敦展廊采用形式设计语言，并根据液体流动形态对建筑外观进行处理，形成一系列建筑元素与自然环境相互作用的动态空间。这里将成为伦敦的一个活动中心，用来举办多样化的活动，如室内/室外展览、会议、演出、研讨会或辩论赛等。

Located at Imperial Wharf close to Chelsea Harbour, the space is not just purely visual, but also uses the art of precision and control to help the visitor understand the relationship between the architecture of the space and the design of Roca bathroom products.

It is her third London project

Using a formal language derived from the movement of water, the Roca London Gallery has been polished by fluidity, generating a sequence of dynamic spaces carved from this fascinating interplay between architecture and nature. It will become a London hub hosting a wide range of activities such as exhibitions produced internally or externally, meetings, presentations, seminars and debates.

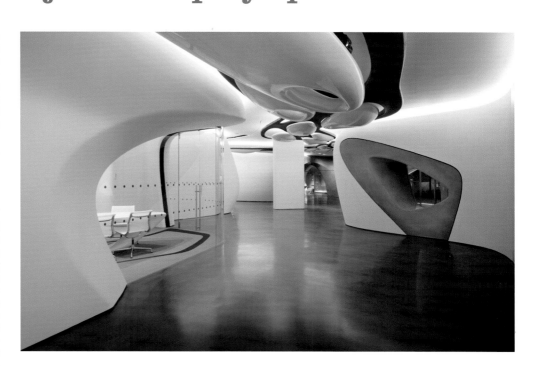

平面图 plan

盒子＝空间
Boxes=Spaces

Crossboundaries Architects
"Siemens Paradigm Shift Office",
Beijing, China
"西门子范式转换办公室"，北京，中国

西门子范式转换办公室对视觉空间的传统习惯提出了非常大的挑战。设计者将办公室的内部空间设计成一个城市环境，提供不同的视角，并且创造出多个狭窄的开放空间，像北京胡同一样拥有丰富的漫步空间。"盒子"设计理念帮我们顺利达到设计目标。

这里共有六个盒子，每个盒子分别提供一种独特的环境。

办公室应该是一个多种行为方式共存的空间

采用玻璃外壳结构实现内部空间在视觉上的通透性，使透明度成为盒子建筑的一个固有特性。

The office is a big challenge to the conventional habits of perceiving space. They treated the interior of the office as an urban environment, providing different points of view, creating open and narrow areas, to enrich the experience of the walk, in the same way that the Beijing Hutong does. The box is the tool that helped us reach our goals.
There are a total of six boxes, each of them offering a unique environment.

An office should be a place where many diverse behaviors coexist

It is the glass enclosure that makes the interior visually accessible, making transparency an inherent characteristic of the box.

前沿科技
Pioneer technology

**Samuel Ruiz Torres de Carvalho
Pedro Palmero Cabezas**
"Business Incubator Centre",
Almada, Portugal
"商务孵化中心",阿尔马达,
葡萄牙

和1999年至2001年间激增的、后来又消亡的所谓"互联网孵化器"一样,"商务孵化器"一词在媒体中流行起来,但商务孵化模型在20世纪50年代就已经出现了。商务孵化器的主要目标是支持商务行为的启动和发展,提供可承受的商务空间以及商务支持和指导。

The term "business incubator" gained popularity in the media with the explosion and subsequent demise of so-called "Internet incubators" between 1999 and 2001, but the business incubation model traces its beginnings to the late 1950s. Its prime objective is to support start-up and growing businesses, offering affordable space and intensive support and mentoring.

全球共有大约5 000家商务孵化器建筑

There are about 5,000 business incubators worldwide

这种建筑形式采用模块化的结构,根据地块的不规则条件和不同功能要求进行削减。这种建筑体量使室内外空间之间的界限变得模糊。

The form of the building is the result of a process in which the mass is moulded and subtracted in response to the irregularity of the plot and the different functional requirements. This dissolution of the volume blurs the limit between exterior and interior.

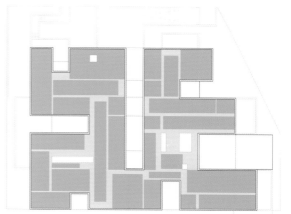
三层平面图 second floor plan

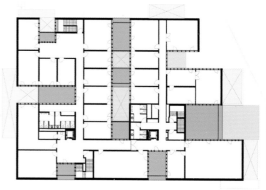
二层平面图 first floor plan

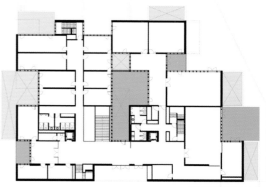
底层平面图 ground floor plan

地下一层平面图 basement floor plan

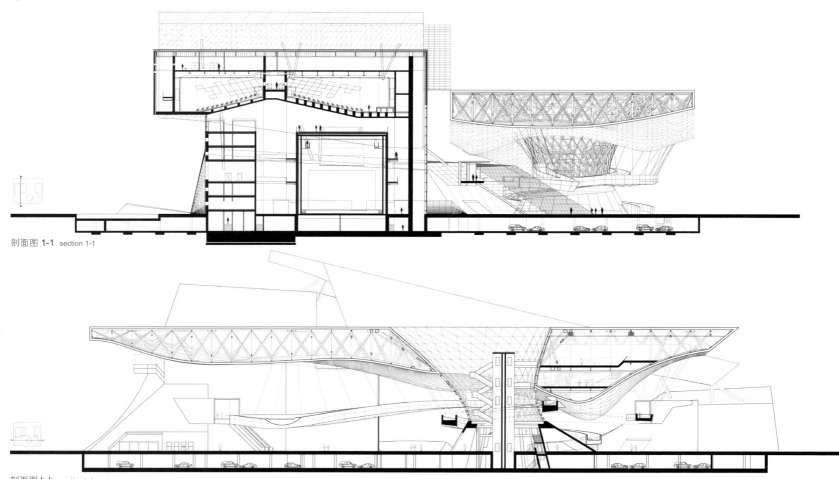

剖面图 1-1 section 1-1

剖面图 b-b section b-b

多功能城市广场
A multifunctional urban plaza

COOP HIMMELB(L)AU
"Busan Cinema Center",
Busan, Korea
"釜山电影中心",釜山,韩国

区块位置 site plan

釜山电影中心项目的基本理念是讨论开放空间与封闭空间以及公共空间与私人空间的堆叠设计。项目位于海云台区U-dong新世界,它将成为亚洲最大的电影活动之家。

The basic concept of this project is the discourse about the overlapping of open and closed spaces and of public and private areas. Located in U-dong Centum City, Haeundae District, the building will serve as the largest home of Asia's cinema events.

我们的建筑目标是克服重力作用

Our architectural aim is to defy gravity

该高层建筑采用世界最大面积的屋顶,空间无柱屋顶最大程度地实现"飞翔的屋顶"的设计理念。独特的照明系统可以根据特定事件进行调整,照射在天花板上进一步增强动态效果。

The multi-story center boasts the world's largest roof, as a column-free roof covering a space comes closest to the idea of a "flyingroof". Adding to the effect is an elaborate lighting system that can be tailored to events and displayed across the ceiling in full motion graphics.

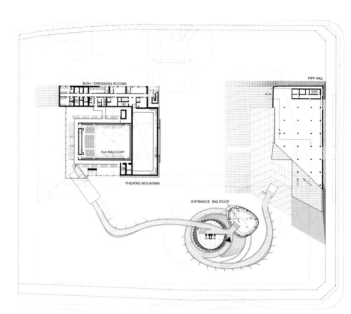

平面图 4 floor plan 4

平面图 3 floor plan 3

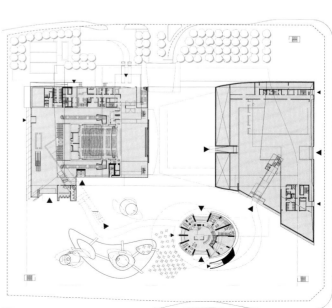

平面图 1 floor plan 1

灯光与色彩
Light and colour

LOVE architecture and urbanism
"Shopping Center Gerngross",
Wien, Austria
"Gerngross购物中心",维也纳
奥地利

Graz建筑公司负责Gerngross购物中心的翻新工程。该购物中心的通道设计混乱复杂,不利于辨识。为了改善这一状况,设计师对室内空间布局进行了重新设计,增加了许多小型零售空间。中庭设有自动扶梯网络,这里是建筑的核心部分,连接水平和竖直方向的视觉轴,而且已经成为该购物中心的中心定位点。

立面采用大面积无定形彩色区域

采用大面积铝材和玻璃材料对主要入口和飘窗的地面装饰进行升级。立面在不同视角呈现不同的华丽外观。此外,整个立面采用热学方法和现代化技术进行处理,使立面的装饰图案看起来像是处于太阳的保护之中。

Graz architectural firm was given the task of refurbishing the Gerngross. Walkways were complex and confusing, and it was difficult to navigate. To improve this, the layout of the indoor levels has been entirely re-designed and smaller retail spaces have been added. The atrium, with its escalator network which unites the horizontal and vertical visual axes, serves as the nucleus and has become the store's central orientation point.

Large-scale, amorphous colour fields were applied to the façade

The floor space at the main entrance and the bay windows has also been upgraded through a large-scale aluminum-glass construction. The appearance of the façade changes, depending upon where the viewer is standing. Furthermore, the entire façade was thermally reengineered and technically modernized, so the ornamental pattern functions simultaneously as sun protection.

模仿水流形态
A simmulation of water flow

KDG group, INC - Kalarch Shanghai
"Shanghai-Zhangpu Aquatic Center",Kunshan, Shanghai, China
"上海-张浦水上运动中心",
昆山,上海,中国

受昆山政府委托建设该项目,该建筑综合体包括三个不同的游泳区域。建筑立面采用U形玻璃嵌板进行隔热,并且创造出一种柔美光亮的波纹状皮肤效果。

建筑外观模仿水流的形状

建筑师在外观与内部空间、有机形状和直线形式之间造成鲜明对比,使我们感觉到水的特征。该建筑并不是一个简单的圆桶,每一楼层之间都有相对位移,避免立面形成垂直单一的线条。

Commissioned by the Kunshan Municipal Government, the complex houses three different swimming areas. The façade is designed with U-profiled glass panels that act as good thermal insulation and create a soft, light and wavy skin.

The appearance mimics the flow of the water

The contrast between the envelope and the content, the organic shape and the straight form, is the best way of the architects to give our tribute to water. The building is not a simple cylinder, and the different floors are displaced in each level, avoiding to form vertical line of the façade.

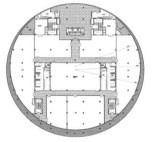
二层平面图 first floor plan

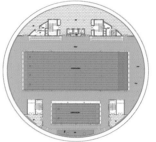
四层平面图 third floor plan

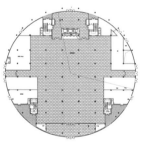
底层平面图 ground floor plan

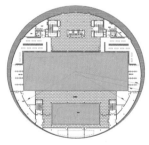
三层平面图 second floor plan

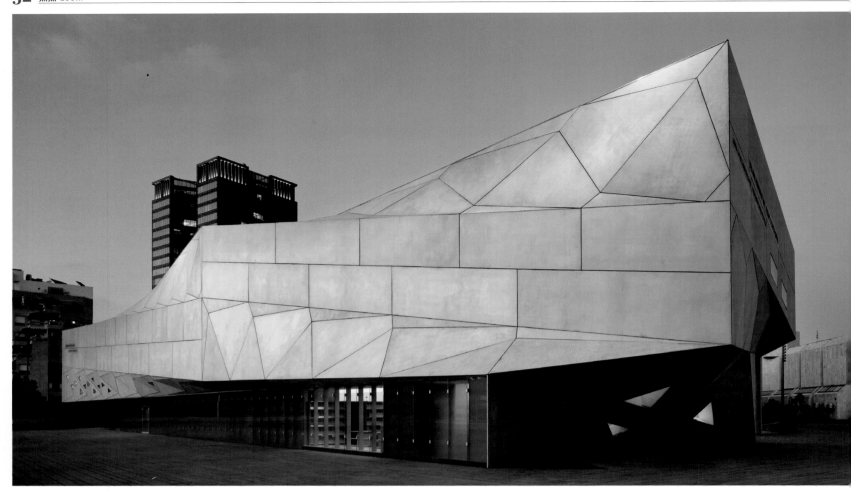

博物馆新体验
A new type of museum experience

PRESTON SCOTT COHEN, INC.
"Tel Aviv Museum of Art", Israel
"特拉维夫艺术博物馆", 以色列

本项目的设计工作面临的挑战是在一个有限的特殊三角形地块上建造多层建筑, 提供大面积中性矩形画廊。解决方案是根据不同轴线使三角形的水平空间成为方形空间, 每层之间的轴线具有较大偏差, 通过一个约26.5米高的螺旋状中庭将各平面统一起来, 在中庭顶部设置灯光照明, 在建筑内部形成精美的弯曲表面。该建筑成功整合了当代艺术博物馆看上去无法调和的两种模式, 即中性白色盒子建筑和展览博物馆。

相互堆叠起来的独立平面

The design for the Amir Building arises directly from the challenge of providing several floors of large, neutral, rectangular galleries within a tight, idiosyncratic, triangular site. The solution is to "square the triangle" by constructing the levels on different axes, which deviate significantly from floor to floor. The levels are unified by an 26.5-meter-high, spiraling, top-lit atrium, whose form is defined by subtly twisting surfaces that

Independent planes stacked one on top of the other

curve and veer up and down through the building. The building combines two seemingly irreconcilable paradigms of the contemporary art museum: the museum of neutral white boxes and the museum of spectacle.

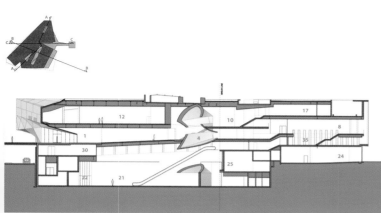

剖面图 a section a

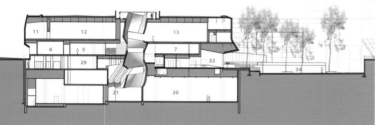

剖面图 b section b

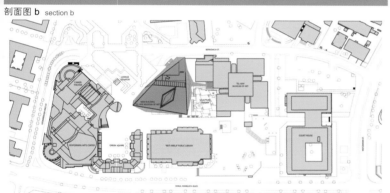

区块位置 site plan

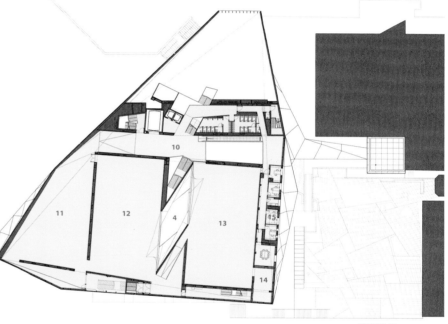

平面图 02 floor plan 02

平面图 01 floor plan 01

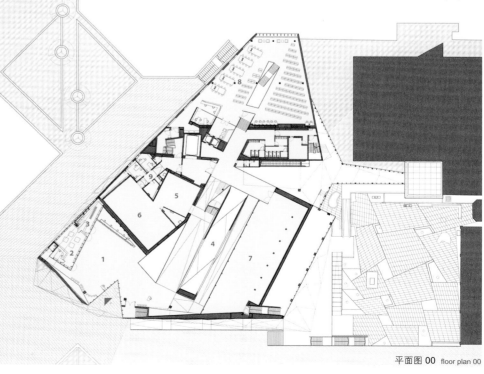

平面图 00 floor plan 00

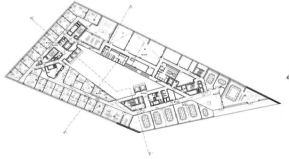
二层平面图 first floor plan

三层平面图 second floor plan

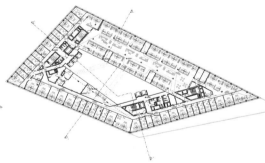
十四层平面图 13th floor plan

Spiegel总部大楼　Spiegel Headquarters

Henning Larsen Architects
"Spiegel Group Headquarters", Hamburg, Germany
"Spiegel 集团总部"，汉堡，德国

这座新建办公楼是这个港口城市HafenCity海港区的一个重要部分，HafenCity海港区目前是欧洲最大的市内开发项目。该项目包括两个建筑，形成一个巨大的U形格局，拥抱着所面对的城市空间。两个建筑之间形成两个广场，分别是一个面对Brooktorkai区供步行者、骑车者和机动车使用的入口广场和一个与滨海散步道直接相连的开放式公共广场。

建筑共13层，窗户采用白色金属边框进行装饰

The new office building is an important part of the port city's ambitious HafenCity, currently Europe's biggest inner-city development project. The two buildings are designed as large U-forms that embrace the urban space they are directed

The building is comprised of 13 stories, and the windows are framed with white metal

towards. The two buildings form two plazas: an arrival plaza for pedestrians, cyclists and drivers towards Brooktorkai and an open public plaza, which has a direct connection to the waterfront promenade.

底层平面图 ground floor plan

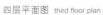
四层平面图 third floor plan

十层平面图 ninth floor plan

Ericus-Contor大楼 Ericus-Contor Building

Spiegel总部乔迁新址
Spiegel moves to a new place

绿色城市皮肤
Green and urban skin

Casanova + Hernandez Architecten
"Ginkgo Apartments Building", Bergbeek, Apeldoorn, The Netherlands
"银杏公寓大厦",伯格比克,阿珀尔多伦,荷兰

银杏公寓大厦项目位于荷兰Veluwe自然公园附近,俯瞰一个旧式教堂和该镇的中央公园。该项目试图探索为不同目标人群提供经济适用房的可能性,采用紧凑的住宅建筑综合体,从物理结构和视觉效果上与周边环境融为一体。

The project is located near the natural park of Veluwe in the Netherlands with views over an old church and the central park of the small town. Ginkgo project explores the possibilities of providing affordable housing for different target groups by using a compact housing complex, physically and visually integrated in its context.

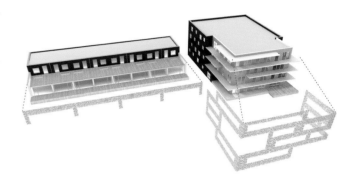

使艺术、科技和建筑特点相结合
It combines art, technology and architecture

本项目的设计重点是两个外墙之间的对话以及与场所两边环境之间的相互协调。

The project is based on a careful dialogue between two skins that give answer to the border conditions on both sides of the location.

YVR, Lysbeth de Groot-de Vries
"Busstop Park+Ride Citybus", Hoogkerk, The Netherlands
"停车场与公交换乘站",霍克尔克,荷兰

三角形广场采用百合花瓣造型,该造型延伸至建筑内部。建筑的主结构为一个三面开口的圆顶混凝土壳体。

The triangular square is characterized by a pattern of lily leaves that extends to inside the building. The main structure of the building is a domed concrete shell broken on three sides.

建筑外形是一个鲜明的雕塑景观

The shape of the building is a clear striking sculpture in the landscape

无直角建筑
No right angles

马德里竞技提出新足球场规划方案
Atlético Madrid presents new stadium plans

Cruz y Ortiz arquitectos
"Football Stadium Atlético de Madrid", Madrid, Spain
"马德里竞技足球场",
马德里，西班牙

马德里竞技中心公布了其在马德里东部建造新球场的计划。新球场将容纳67 500人，比目前球场的容量多12 000人。目前的球场位于一条河流、一条高速公路和众多住宅之间，没有扩建的可能性。

**当前赛程安排
期望新球场于
2015年竣工**

新球场将建在目前的Estadio La Peineta球场位置，该球场目前主要用于田径运动。如果马德里获得2020年或2024年奥运会的主办权，该球场将作为比赛的主要场地。看台上方设计了一个灯罩，就像一个巨大的斗篷，保护球场的观众，并与球场形成一个整体。

Atlético Madrid unveiled its plans to build a new stadium in the east of Madrid. The stadium will have a capacity of 67,500 seats, about 12,000 more than the current stadium. The current stadium lacks the possibilities for expansion being located in between a river, a motorway, and residential buildings.

The current timetable expects the stadium to be completed in 2015

The stadium will be built at the site of the current Estadio La Peineta, which is currently mainly used for athletics events. If Madrid gets awarded the right to host either the 2020 or 2024 Olympic Games, the stadium will serve as the main venue for the games. A light cover, which protects the spectators, sits above the stands as if it were a large mantle, providing the intervention with unity.

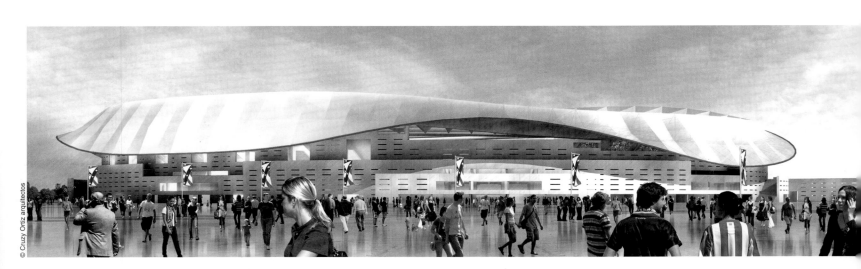

DNA Barcelona Architects
"Le Far du Grand Vent",
Cheraga, Algeria
"远航大厦",
切拉贾, 阿尔及利亚

远航大厦项目与城市环境相互协调，并且与周围环境良好地融合在一起。建筑外观采用一种未来派破冰前行的帆船造型。该建筑已经成为该地区城市天际线中的一个地标性建筑。在夜晚，灯光穿过巨大的玻璃表面，使建筑成为一个光点，产生更加迷人的效果。

建筑造型从自然中获得灵感

公寓和酒店拥有优美的景观视野。建筑设有大型露台，创造出面对城市空间的私人空间。较低楼层提供公共活动空间。

This project adapts itself to the surroundings. It follows the forms of the urban surface. The building itself reminds of a futuristic ship, melting ice and going forward. It becomes a landmark at the city-skyline. During the night, it becomes even more attractive and turns into a luminous spot, as the huge glass surfaces let the light pass through.

Nature as an inspiration for the forms of the building

The apartments and hotel have privileged views over the landscape. There are also big terraces to create private promenades towards the city. The lower floors house the public programs.

城市中的自然元素
A natural element in the city

立方体建筑
A cube-shaped building

Mecanoo
"Hilton Schiphol Airport", Amsterdam, The Netherlands
"希尔顿史基浦机场酒店",阿姆斯特丹,荷兰

希尔顿史基浦机场酒店将取代现在的希尔顿地产,项目计划于2012年末开始施工,并于2015年开业。

全新的旅店体验是其追求的目标

1959年,倡导机场酒店概念的希尔顿集团于旧金山机场建设的希尔顿旧金山机场酒店开业。目前其公司在全球范围内已经拥有320多家机场酒店,是机场酒店行业的领军企业。希尔顿正在寻找机会进一步增加其在全球运输枢纽地区的酒店数量。

The new hotel will replace the current Hilton property and is expected to open in 2015, with construction scheduled to start later in 2012.

A completely new hotel experience is set to become a destination in its own right

Hilton pioneered the airport hotel concept with the opening of Hilton San Francisco Airport in 1959 and, with the company now having more than 320 airport hotels worldwide, they are further seeking to increase the number of hotels at the world's leading transport hubs.

0720 Fermín Vázquez Arquitectos + Jaime Lerner Arquitetos Associados
"Waterfront of Port of Porto Alegre", Brazil
"阿雷格里滨海港口",巴西

阿雷格里滨海港口项目的目的是对所谓的Cais Maua码头进行改造,将其从目前的废弃和衰败的码头区改造成一个商业、文化和娱乐中心,为2014年世界杯提供服务。

The project aims to transform the so-called Cais Maua, currently disused and decayed, into a center of business, culture and entertainment for the World Cup 2014.

本项目计划将耗资大约1.98亿欧元

The project is expected to cost approximately €198 million

本项目占地18.1万平方米,根据可持续性标准进行规划,将包括一个商业园区、一个酒店和一个购物中心。

The 181,000 m² project will house a business park, a hotel and a shopping center planned under sustainability criteria.

尊重文化的重要性
Significance of respecting culture

Architrend Architecture (Gaetano Manganello & Carmelo Tumino)
"Architrend Office", Ragusa, Sicilia, Italy
"Architrend办公楼",拉古萨,西西里,意大利

该建筑采用钢筋混凝土结构,并以玻璃覆盖,是现代建筑的一个典型案例,具有新巴洛克式、新摩尔式、西班牙和葡萄牙建筑特点。宽大的外挑结构保护建筑四周的窗户免受太阳光的强烈照射。

最大化利用太阳能辐射

整个建筑的主要构件、玻璃、人行道、内墙和家具全部采用边长为120厘米的模块化网格。

Made of reinforced concrete and covered in glass, this building is an example of modern architecture, with neo-baroque, neo-Moresque, Spanish and Portuguese influences. Wide overhangs protect the entire perimeter of the building's windows from the sun.

Maximizing the contribution of solar radiation

A modular grid of 120cm rules the main components of the entire construction, glasses, pavements, interior walls and furniture.

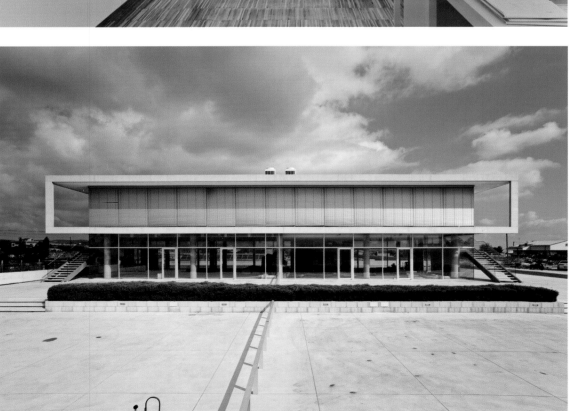

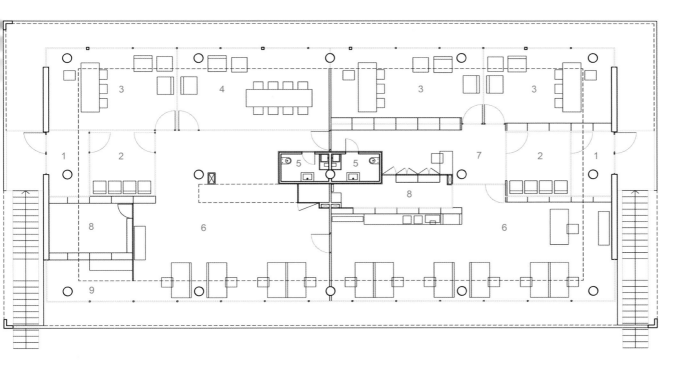

巨大的框架
A large frame

住宅
RESIDENTIAL

下层平面图 lower floor plan

上层平面图 upper floor plan

两个建筑体量的交汇
The intersection of the 2 masses

VaSLab Architecture
"Baan Bueng Residence", Cholburi, Thailand
"Baan Bueng住宅"，春武里，泰国

Baan Bueng住宅项目采用紧凑布局，位于距离曼谷东部90千米的春武里府Baan Bueng。业主委托建筑设计师为他及未来的妻子设计一个半泰国半日本风格的爱巢。底层的轴线主要是白天活动的空间，而上层空间中包括家庭房和主卧室，以及为将来房屋成员设计的一个小卧室。

This compact house is located in Baan Bueng, Cholburi province, 90 kilometers east of Bangkok. The owner asked the architect to design a love nest for him and his future wife, with half Thai and half Japanese style. The lower level axis becomes mainly daytime activities space while the upper level consists of a family room, a master bedroom, and a small bedroom for the future member of the family.

建筑体块的叠加
Superimposition of volumes

G2 Estudio
"Ribbon House", San Carlos de Bariloche, Argentinean
"丝带房屋"，圣卡洛斯德巴里洛切，阿根廷

丝带房屋是两个塔希提岛家庭的度假之地，这两个家庭希望能够在一个梦幻般的环境中共同消磨空闲时间。建筑结构由多个体块堆叠而成，这些体块将整个建筑空间分隔成不同功能、不同外观的多个空间。建筑立面采用石木材料进行装饰。

The residence is the vacation destination for two Tahitian families who want to be able to spend their free time together in a wonderful setting. The juxtaposition of volumes that compose the structure of the house allows the space to be divided into different spaces with different functions and different appearances. The house is wrapped in a stone and wood façade.

> 我们可以欣赏上下空间之间的连接设计
>
> *We can appreciate an up-down experience link*

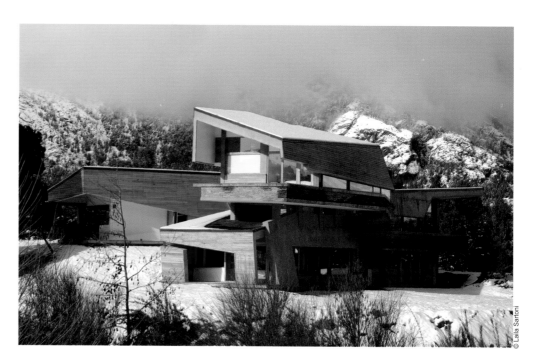

底层平面图 ground floor plan 二层平面图 first floor plan 套房层平面图 suite floor plan

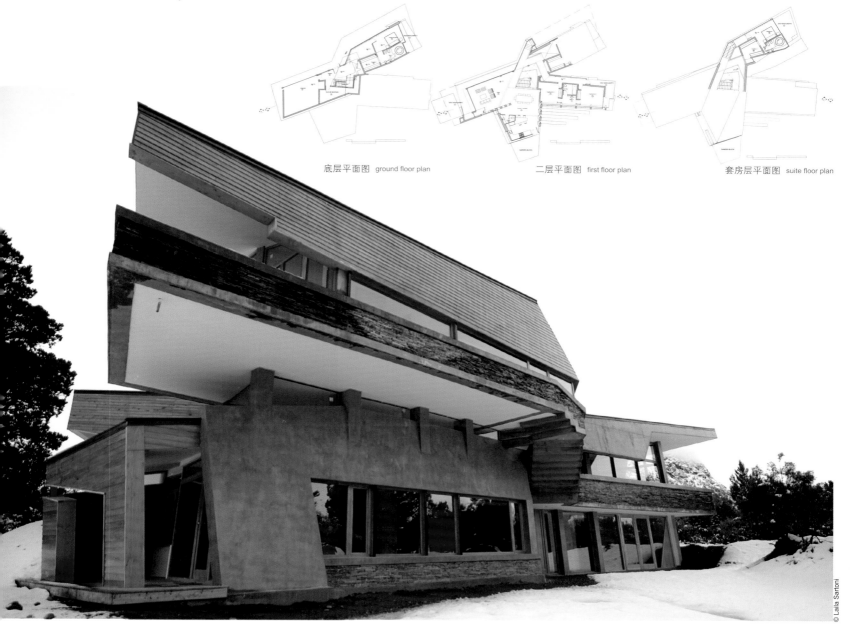

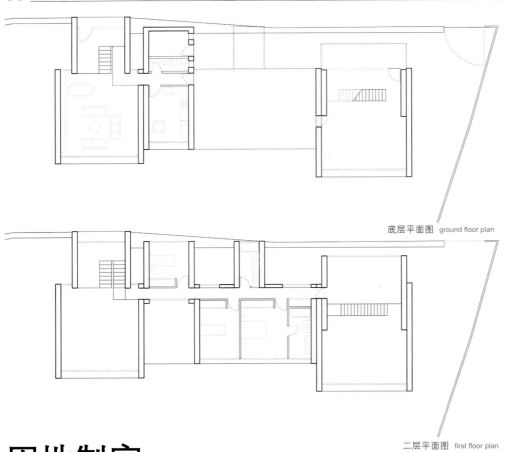

底层平面图 ground floor plan

二层平面图 first floor plan

因地制宜
An environment-concerned treatment

BmasC
"HUETE HOUSE", Muñopepe (Ávila), Spain
"HUETE住宅", Muñopepe(阿维拉), 西班牙

该建筑坐落在山脚与村庄之间的边界上，设计师试图用最少的成本将其改造成住宅建筑。该项目将经济成本作为首要条件，并且优先采用可回收环保材料。

The house is located between the foothill and the village and is planned to be reconstructed to a residence at minimum cost. The project takes economy as a priority, taking advantage of E-Waste recycling materials.

南立面图 south elevation

北立面图 north elevation

创新实验室
An innovative laboratory

地块位置 site plan

3XN and GXN
"Green Solution House",
Rønne, Denmark
"绿色解决方案住宅",
伦讷,丹麦

"丹麦30%的垃圾废物来自建筑行业。"这个扇形建筑完全采用可回收或可降解材料建造。因此,该建筑的设计理念是采用可降解或可回收材料来消除建筑垃圾。

"In Denmark, the construction industry alone is responsible for 30 percent of all waste generated." All materials used in the fan-shaped building are either fully recyclable or biodegradable. Hence, it is a building design that takes on the ambition to eliminate the waste.

设计成为一个最高水平可持续发展的平台

Designed to be a platform for the highest level of sustainable development

底层平面图 ground floor plan

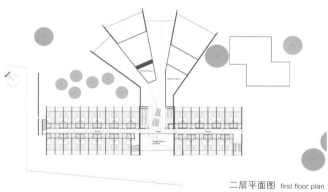

二层平面图 first floor plan

生物材料 bio materials

科技材料 techno materials

水 water

能源 energy

生物多样性 biodiversity

清洁交通 clean mobility

强调感官功能
Accentuating sensations

Edward Ogosta Architecture
"Four Eyes House", Coachella Valley, California, USA
"四眼住宅",科切拉峡谷,加利福尼亚,美国

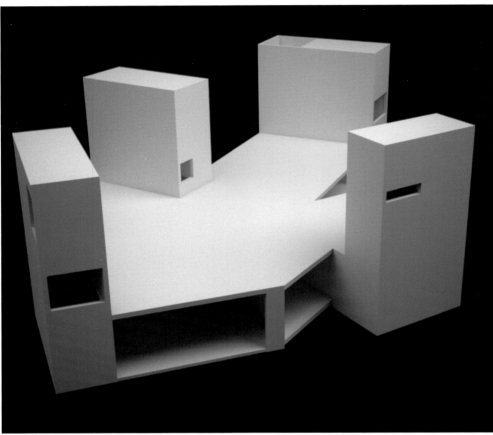

该项目并没有按照传统住宅模式进行规划,而是用其来增强地块上的大量环境元素。"沉睡的塔楼"分别朝向拥有不同视觉体验的四个方向:东侧在早晨可以欣赏日出、南侧面对层层的山脉、西侧在傍晚时刻面对城市绚丽的灯光、上空在夜间悬挂着满天的星星。

Rather than planning the house according to a domestic functional program, architects designed the building foremost as an instrument for intensifying a number of onsite phenomenal events. Four "sleeping towers" are oriented toward four spatiotemporal viewing experiences: morning sunrise to the east, mountain range to the

所有卧室采用相同的尺寸设计

All bedrooms are equally-sized

south, evening city lights to the west, and nighttime stars overhead.

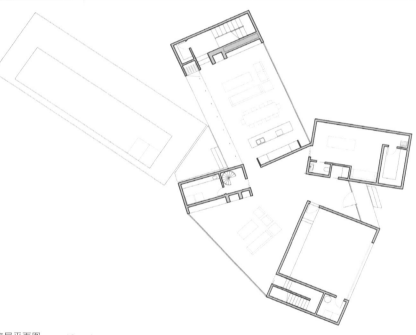

底层平面图 ground floor plan

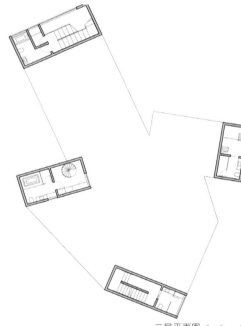

二层平面图 first floor plan

光滑的线条
Sleek lines

Estudio Arquitectura Hago
"RG House",
Don Benito, Badajoz, Spain
"RG住宅",唐贝尼托,
巴达霍斯,西班牙

用更少的元素实现更多的功能是伟大设计的一般规律。该住宅项目的设计目标是解决一个大家庭的住房难题。居住问题和住宅需求的出现是由于该大户人家的成员都住在一起。因此,设计理念是创造一个不同的公共空间(独特而具有识别性)作为其他模块化房屋(卧室、浴室和厨房)的补充,以此来完善该住宅项目。

As is often the case with great design—less is more. The house was built in order to solve the difficulties of a large family. Questions and needs arise from the way in which family members in this large family should live together. The idea is to create a different communal space (special and recognizable) as a complement for the rest of the modular units (bedrooms, bathrooms, kitchen) to complete the program.

实现最小化住宅理念

It celebrates minimalism in architecture

建筑含蓄简洁的外观造型与周边的工业环境互相协调。隐藏式的窗户设计使住宅建筑与周围花园、庭院和水池良好地融合在一起。

The introverted nature of the house, due to its location in an industrial area, inspires its clean shape. The aim of hiding the windows emphasizes how the central space blends with the surrounding garden, courtyard and pool.

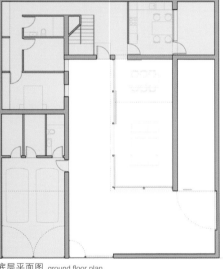
底层平面图 ground floor plan

二层平面图 first floor plan

悠远僻静的建筑
Separate structures

SIMON LAWS
"The Drew House", Gladstone, Australia
"德鲁的住宅",格拉德斯通,澳大利亚

"德鲁的住宅"是澳大利亚著名摄影艺术家玛丽安·德鲁(澳大利亚最具潜力以及最重要的影像媒体艺术家之一)及其家人度假的地方。

地面上建造了预制的抛物线形屋顶结构和多块甲板

该住宅建筑的设计目的是在临近海洋的红木林和古老棕榈树林之间创造一种奢华的露营地,建筑形式应该充分反映这一家庭曾涉足制糖业的悠久历史。与住宅相互分离的凉棚为度假家庭提供私密空间,其完美地融入到周围的环境当中,并对环境产生最小的影响。

The Drew House is a holiday retreat for Australian photographic artist, Marian Drew (one of Australia's most influential and significant photo-media artists), and her family.

The prefabricated parabolic roofed structure and decks are erected onsite

The house aims to create a kind of luxury campsite nestled amongst the Bloodwoods and ancient Palms near the ocean with forms that also reflect the strong iconography of the sugar mills that were also a large part of the family's history. The separated pavilions allow for the privacy of holidaying family groups and also more easily integrate into existing landscape, which has been minimally disturbed.

MAXXI
"OFFICINA ROMA",
Rome, Italy
"罗马办公室",
罗马，意大利

"罗马办公室"包括一间卧室、一间厨房和一个工作间，由来自意大利各地的24名高中生耗时一周的时间建造完成。厨房完全采用废旧瓶子建造，卧室的墙壁是报废车辆的车门，工作间采用木窗和旧家具建造。

材料拼接建筑

该建筑是专门为可回收材料作品展(RE-CYCLE)设计的参展作品，该展览在罗马MAXXI博物馆持续5个月的时间。该建筑基于自然资源的特征、完善(竞争)、增长以及开发，讨论并质疑人类生活方式的必要性。

It consists of a sleeping room, a kitchen and a work shop. It has been built in one week by 24 high school students from all over Italy. The kitchen is entirely built out of old bottles, the sleeping room walls out of used car doors, and the workshop out of wooden windows and old furniture.

The building is a collage of materials

It was specially created for the RE-CYCLE exhibition, which is running for a period of 5 months at the MAXXI in Rome. It talks about the essential necessity to question our lifestyle, based on individuality, completion (competition), growth and exploitation of natural resourses.

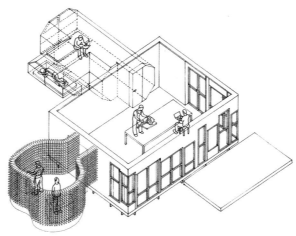

平面图 plan

完全采用废弃物建造
Entirely built out of trash

盒子状几何体
Boxed geometries

五层平面图 fourth floor plan

Kavellaris Urban Design
"Jewell", Melbourne, Australia
"Jewell住宅",
墨尔本，澳大利亚

该住宅的南北两个侧面采用不对称的平滑拐角设计，将收敛的建筑立面融入延续的建筑外观中，产生一种动态效果。

采用变化的纹理与颜色

本案结合地块内的景观街道和小巷在每户之间提供一定的间隔，使每个可居住的空间都能够获得良好的自然光和通风效果。从不同角度观看，该建筑可以有多种不同的视觉感受。

The asymmetrical composition of slipping corners that wrap around the northern and southern edges of the building blurs the boundaries from the converging facades into a continuous architectural expression that evokes a sense of movement.

The textural qualities and colour of the building change

The project incorporates landscaped streets and laneways within the subject site to provide spatial separation between dwellings, enabling every habitable space to receive natural light and ventilation. As the building is experienced from the various vantage points, the building form takes on different readings.

二层平面图 first floor plan

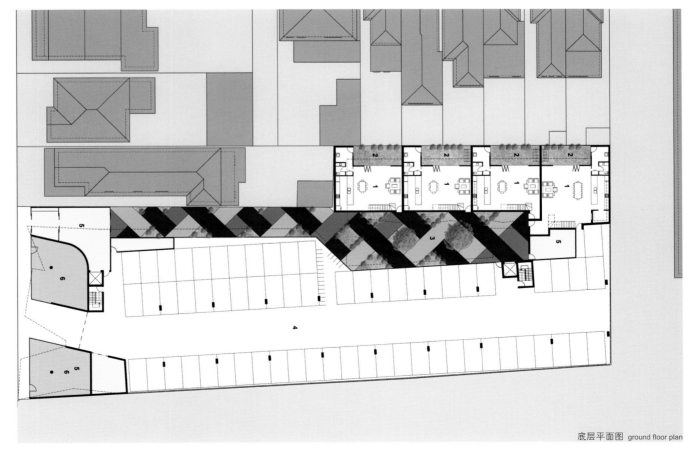

底层平面图 ground floor plan

Pichler & Traupmann Architekten
"Haus P-K", Punitz, Austria
"P-K房屋"，普尼茨，奥地利

区块位置 site plan

具有农业特色的建筑
An agricultural character

该建筑的两侧都是堤坝，堤坝两翼是护堤墙。作为一种建筑元素，护堤墙从地面起升至楼层的高度，面对主要入口。入口的水平位置几乎完全嵌入到地形当中。

沐浴在日光中的空间

上部空间是一个完全开放的空间，其中包括私人居住空间。室外空间通过几个大露台相互连接，在靠近房屋的露台上面设计了多处花园景观。

It starts with an embankment on both sides and is then flanked by a retaining wall that, as a building element, rises continuously out of the ground and reaches storey height towards the entrance. The entrance level is almost completely embedded in the terrain.

Spaces that are bathed in daylight

The upper level with the private living areas is set completely free. The outdoor spaces are connected through big terraces, which are designed as gardens when they are close to the house.

二层平面图 first floor plan

底层平面图 ground floor plan

地下一层平面图 basement floor plan

西班牙对经济适用房仍然具有需求
Spain still demands affordable housing

VOLUAR ARQUITECTURA
(Borja Lomas Rodríguez+Pablo Rodríguez Mesa)
"368 Social Housing", Getafe, Madrid, Spain
"368户社会住房",赫塔菲,马德里,西班牙

该项目反映了"封闭街区"概念以及建筑空间与用户之间的关系。建筑立面采用多样化的彩色垂直金属条纹进行装饰,为城市提供了一种现代化的建筑形象。这些建筑拥有两种立面效果,一种采用暖色调,另外一种采用冷色调。两种色调之间通过一个中性结构实现完美过渡。

The generation of the project is a reflection on the concept of "closed block" and the relationship between the spaces generated and building users. It is presented to the city with a modern look, variable, shaped with vertical metallic stripes of different colours. The buildings have two faces, one with warm colours, and the other with cool colours, both articulated by a neutral body which resolves the transition between these façades.

提供"住房"还是提供"社会福利"?
Providing "housing" or providing "social welfare"?

额外的楼层
An extra floor

BYTR Architects
"Lighthouse", Utrecht, The Netherlands
"灯塔住宅",乌特勒支,荷兰

作为砖房上的额外扩建部分,项目的设计理念是使该结构尽可能轻,因此采用了镶嵌着铝板的木材进行建造。有趣的立面设计巧妙地展示了奇妙的灯光效果:晚上,金属从后面被照亮;白天,日光穿过小孔进入室内。

Built as an extension above a brick house, the structure was made as light as possible using timber lined with aluminum panels. The interesting feature of facade design is magic play of light: at night, the metal is lit from the back, while in the day, the perforations provide diffuse daylight into the interior.

二层平面图 first floor plan   三层平面图 second floor plan

屋顶设置了绿色花园
Roofs are designed with green gardens

LR arquitectos – Gilberto L. Rodríguez
"BC House", Monterrey, Nuevo León, Mexico
"BC住宅"，蒙特雷，新勒昂，墨西哥

该住宅拥有极好的景观，南面是Chipinque国家公园、东面对蒙特雷市边界上独特的Cerro de la Silla山峰。建筑采用简洁纯色几何结构和极具挑战性的结构设计，从而在沉重且庞大的建筑体量中展示一种灯光效果。

The house enjoys excellent views towards the south with the National Park of Chipinque, and towards the east, dominated by the Cerro de la Silla, an emblematic hill in the boundaries of the city of Monterrey. With simple, pure geometric volumes, but rather challenging structural solutions, the project intends to evoke an image of lightness within a language of heavy and massive volumes.

黑色矩形建筑采用花岗岩材料

The dark rectangular volumes are made from black granite

建筑采用许多悬空体块，从而创造出一种独特的立面效果和一个开放的内部空间。

The home is comprised of stacks of cantilevered volumes that create a unique façade and an open interior.

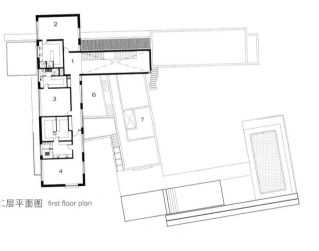

二层平面图 first floor plan

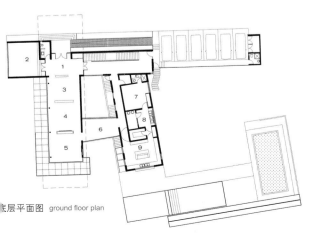

底层平面图 ground floor plan

设计
DESIGN

一款新颖的书架
A trendy bookshelf

Mebrure Oral
"TarGetBookshelf"
"TarGet书架"

这位土耳其工业设计师已经发现了一种在墙壁上展示书籍的当代方法，彻底将书籍从书箱中解放出来。这款书架突出各部分之间的分隔效果，即通过二层设计能够轻松帮助人们找到相关书籍。

This Turkish industrial designer has found a contemporary method of displaying books on walls, making books straight out of the box. The bookshelf features separate sections that help people easy to find books with a two-level design.

您是否想要一款既简单实用又可以对书籍进行有效分类的创意书架？

Do you want a bookshelf that's creative, simple and keeps your books organized?

这款书架的三个部分采用可回收的ABS注塑材料制成。

The three pieces of the bookshelf are made of recyclable ABS plastic using injection moulding.

可移动的卧室
Portable bedroom

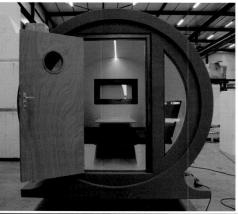

ube4all B.V.
"The Tube"
"管状房屋"

这是一款可以用做不同功能的新型产品，可以放置在任何需要临时休息的场所内，如烟室、工作间、假日公园和海滩等。

A new product which can be used for different purposes can be set in smoking room, cabin, holiday parks and beaches and

该管状房屋的质量为450磅

The weight of the Tube is 450 pound

everywhere where a temporary rest space is desired.

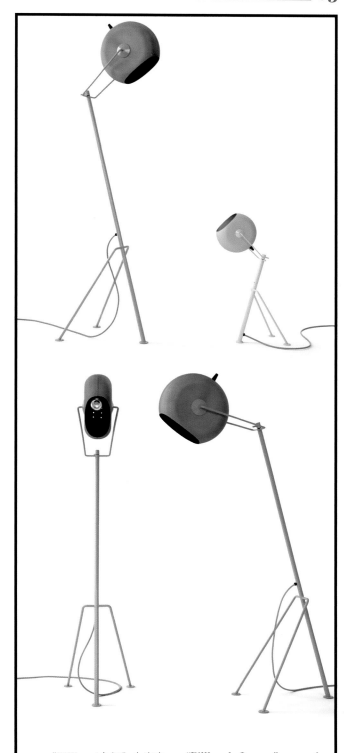

"Pillhead台灯"由来自匈牙利duo a+z设计工作室的设计师Attila Kovacs和Zsuzsa Megyesi设计完成。落地灯和台灯是该工作室"sen-UFO创意"理念系列产品的一部分，该系列产品全部由青色、洋红、黄色和黑色等不同颜色的铝材制成。该理念系列还包括CMYK海龟形小桌系列。

"Pillhead Lamps" are the work of Hungarian duo a+z design, made up by Attila Kovacs and Zsuzsa Megyesi. The floor and table lamps are part of their "senUFO-originals" concept series, in which all products are composed of aluminum and in the colours of cyan, magenta, yellow and black. "SenUFO-originals" also includes a "CMYK Turtles" line of small tables.

是否感觉到工作压力大?
Stressed at work?

有时我们无法逃脱工作中的压力。工作结束期限一个接着一个到来，老板开始大声叫喊，你甚至想要起身离开。设计师Bashko Trybek已经设计出一种可以缓解压力的座椅方案，即由上百个可以调节的压力球组成的座椅。

Sometimes there's no escaping from stress on the job. Deadlines come one on top of the other, and then bosses start yelling, and as much as you'd love to just stand up and leave. Designer Bashko Trybek has designed a way to combat stress with a chair with hundreds of rearrangeable stress balls.

增加你的个人感受
Add your personal touch

根据"线条式"设计理念，该产品采用一个钢管结构和几条线支撑一个漆布桌面。这些线可以变色，能提供丰富的色彩效果。

The design presents "on a string", a structure of tubes and strings which supports a linoleum table top. The string can change its colour.

您是否想将自己从这个世界上隐藏起来？

您是否想仅仅躺在某处就可以将自己从烦恼中掩藏起来？

Do you want to hide from the world?

Do you just want to lie down somewhere
and hide from your troubles?

飞涨的房价和缺少有效空间等因素正在逐渐逼迫人们寻找其他替代产品。设计师 Freyja Sewell 设计了一种HUSH封闭空间，为用户提供一个私人隐居空间、一个虽黑暗但安静自然的空间或一种安静的心情。

Soaring property prices and lack of available space are causing more and more people to seek alternatives. By creating an enclosed space HUSH, designed by Freyja Sewell, it provides a personal retreat, an escape into a dark, quiet, natural space, or a state of mind.

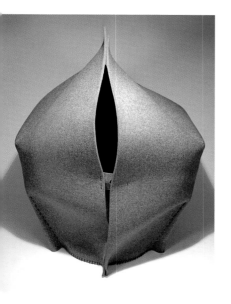
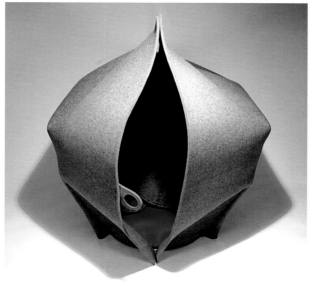

最大化多样性
Maximum versatility

Monoatelier
"Social Noise Office", Milan, Italy
"社会调查办公室",米兰,意大利

该项目是社会调查组织在意大利的第一间办公室,社会调查组织是马德里市内一个社会媒体交流与营销平台。

The project is the first office of Social Noise in Italy, a platform for communication and marketing in social media, based in Madrid.

对材料和空间解决方案进行调查研究的良好机会

A good opportunity to do research on materials and spatial solutions

空间的羊毛窗帘采用螺旋设计,由一个管状钢结构支撑,将不规则的外墙和现有的办公家具巧妙地隐藏起来。

The spiral is used to shape the space, materialized in a wool curtain, supported by a tubular steel structure, which sinuously hides irregularities of external walls and existing furniture.

从蘑菇的形状中获得灵感
Fascinated by mushroom´s biology

DAVID FOX DESIGN LIMITED
"Magic"
"魔法座椅"

魔法座椅从自然界中的蘑菇形状中获得灵感,座椅的座面类似长大的蘑菇,下面的钢制底座提供稳定的支撑效果。

"我喜欢大自然产生的形状,因为它们永远可以达到平衡状态,并且拥有它们各自的平衡比例,这一理念既有趣又可以实现漂亮的外观效果。"大卫·福克斯解释说。

Magic seat draws inspiration from nature's fungi. The shape of the seat takes a mushroom like form, whilst the steel base replicates the delicate channels that sit under the cap.
"I like shapes that nature produces, because they are always well balanced, and promote their own sense of proportion. This concept has both fun and beautiful elements." David Fox said.

日常家具
Everyday furniture

这是一些体积足够紧凑、重量足够轻的实用家具。这些座椅可以根据所用的材质展示不同的颜色和品质效果。

These pieces of useful furniture that are compact and lightweight enough. These chairs made from different materials can display different colour and effect.

全包围式座椅
An all-enveloping armchair

Lou是一款由Patrick Jouin设计的注塑聚酯树脂壳状扶手椅，由软垫，镀铬钢筋和光亮黑漆支腿组成。Lou扶手椅是为巴黎圣奥诺雷街(St Honoré)上的文华东方酒店(Mandarin Oriental Hotel)开业设计的一款家具。

Lou is an armchair designed by Patrick Jouin with injected polyester resin, overlaid and upholstered, resting on a chromed steel or satin black lacquered base. Lou has been designed as furniture for the opening of the Mandarin Oriental Hotel in rue St Honoré, Paris.

Martin Mostböck
"FLAXX Chair"
"FLAXX座椅"

FLAXX座椅是一种混合设计，提供一种自由摆动的舒适享受以及传统的四条腿座椅的功能。

室内外皆适用

座椅的座面采用天然纤维垫制成，经过多层模压形成稳定的立体形状。

The FLAXX Chair is a hybrid design which offers the comfort of a free-swinging chair as well as the functionality of a conventional four-legged chair.

Suitable for both indoor and outdoor use

The seat shell is made from natural fiber mats, which are molded into a stable three-dimensional form by pressing multiple layers of the material.

百分百可回收材料制作
100% recyclable

童年感受
Childhood feelings

Florent Coirier
"Portique Armchair"
"Portique扶手椅"

Portique扶手椅的设计灵感来自花园里的秋千以及Nantes/SaintNazaire港口货物移动装卸台。弯曲的铝管营造了一种魔幻效果。

The chair is inspired by the garden swings, as well as walk-loading of goods from the port of Nantes/SaintNazaire. The aluminum tube is bent creating an illusion.

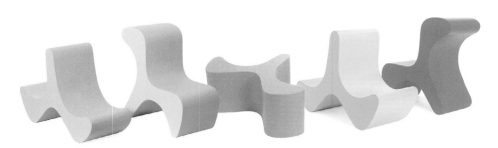

好玩!
Have fun!

Marco Hemmerling
"FLIP chair"
"FLIP座椅"

FLIP座椅采用耐磨软泡沫材料制成，是娱乐休闲的最佳(安全)选择。座椅的座面成弯曲形状，是室内、外空间使用的理想选择。

Made from resistant and cushy foam, the chair is perfect (and safe) for playtime. The seat of the chair is curvy in shape and

共有六种亮丽颜色可供选择

Designed in 6 available cheerful colors

ideal to be used in indoor and outdoor spaces.

五彩坐垫
Multicolored cushions

由克罗地亚设计师Stefan Krivokapić设计的堆叠安乐椅的创意来自使坐垫不停运动的想法，使其就像彼此随意堆叠在一起。

Stack easy chair, designed by Croatian designer Stefan Krivokapić, comes from the idea of having cushions in constant motion that look like they have been stacked randomly on top of each other.

奖项
AWARDS

2011年度詹克斯奖——埃里克·欧文·莫斯
Jencks Award 2011 - Eric Owen Moss

每年一次的詹克斯奖授予为建筑理论和实践做出重大贡献的个人或案例。

埃里克·欧文·莫斯成为2011年度詹克斯奖的获奖者

以往的获奖者包括：2006年扎哈·哈迪德、2007年UNStudio事务所的本·凡·贝克和卡罗琳·伯斯、2008年蓝天组建筑设计事务所(Coop Himmelb(l)au)的Wolf D. Prix、2009年查尔斯·克里亚、2010年史蒂文·霍尔。

The annual Jencks Award given to an individual or practice that has made a major contribution to both the theory and the practice of architecture.

Eric Owen Moss has been 2011 Jencks Award winner

Previous winners include: 2006 Zaha Hadid, 2007 Ben van Berkel and Caroline Bos from UNStudio, 2008 Wolf D. Prix from Coop Himmelb(l)au, 2009 Charles Correa and 2010 Steven Holl.

Samitaur Tower, Culver City, California

AH建筑师事务所在第五届NAN大赛中获得"最佳建筑能量整合方案奖"
AH Asociados receives the 5th NAN Award in the category Best Integration of Energy in Architecture

该建筑师事务所在很长一段时间内一直在西班牙潘普洛纳市寻找一个建设空间,最终他们得到了对一座在开放农村环境中的废弃建筑进行翻新的机会,虽然处于农村环境,但该项目用地与城市和大学具有密切联系。本项目是一个旧锯木厂翻新工程的重要部分,保留当前的结构和外围围护结构,内外空间将采用不同的处理方式。

For a long time the office has been looking for a space within the city of Pamplona, Spain to establish, and finally they came up with the opportunity to refurbish an abandoned building in an open, rural environment but very well connected with the city and the university. The project is set as an integral refurbishment of an old sawmill, maintaining its current structure and peripheral enclosure which is embodied in two different treatments in its interior and exterior.

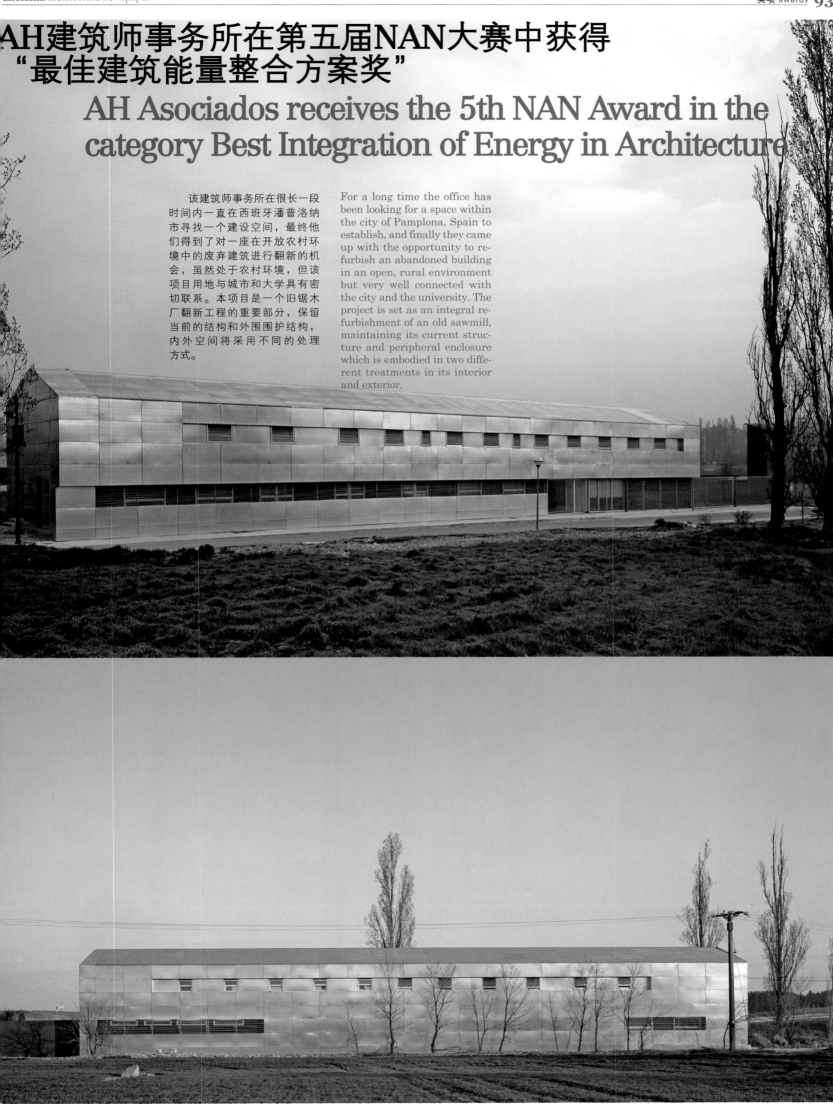

"MIPIM AR未来项目奖"专门为未建或未完工项目设置,共分八个奖项。未来项目奖于2002年在国际房地产市场MIPIM设立,2012年3月6日-9日已于戛纳举行,吸引了2万多名代表参加。

Okay大厦项目设计师Isay Weinfeld获得"2012年整体规划奖"。

The MIPIM Architectural Review Future Projects Awards are for unbuilt or incomplete projects spanning across eight categories. The awards have been running since 2002 and took place at MIPIM, the international property market, which has attracted over 20,000 delegates to Cannes between 6-9 March, 2012.
Okay Building, by Isay Weinfeld is 2012 Overall Planning Winner.

"MIPIM AR未来项目奖"
MIPIM AR Future Project Awards

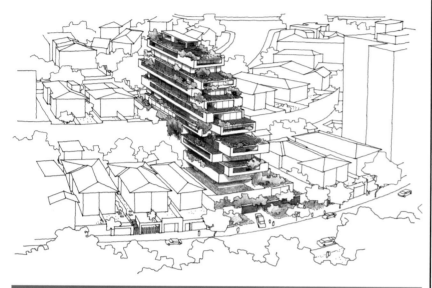

"2011年度A+奖"最佳可持续建筑项
2011 A+ Award
Best Project of Sustainable Architecture

"2011年度A+奖"评选活动期间,观众(由200多名建筑师和专业人士组成)使用一种实时交互式投票系统从评委会预选的项目中评选出每个类别的获奖者。

获奖者:
梅根建筑师事务所(Magen Arquitectos)负责设计的位于西班牙萨拉戈萨Riberas del Ebro市政环境服务总部大厦项目荣获"2011年度A+奖"。

During 2011 Architecture Plus Awards Gala Performance, the same audience (made up of over 200 architects and professionals) chose the winner from each category, with an interactive voting system in real time, from the previous selection made by the jury.
Winner:
Magen Arquitectos have become the winner of 2011 A+ Award for the Headquarters for the Environmental Municipal Services of the Riberas del Ebro in Zaragoza, Spain.

2011年度 "SAIE评选活动"
SAIE Selection 2011

博洛尼亚公司(BolognaFiere)与欧洲建筑(Archi-Europe)网站组织了此次"SAIE评选活动"，活动以"创新、整合、建筑——高创新性和可持续性建筑"为主题，评选出24个获奖建筑项目或设计理念。获胜项目颁奖仪式于"建筑与新生代"论坛期间在SAIE展览会上进行。"青年建筑师组"四个区域的获奖者包括：
des+Bettini建筑师事务所，意大利，木质结构
HArquitectes+dataAE建筑师事务所，西班牙，混凝土结构
LAN建筑师事务所，法国，砖结构
CAFè建筑师事务所，意大利，金属与玻璃结构

BolognaFiere and Archi-Europe have organised the competition entitled SAIE Selection with the aim of selecting 24 projects or ideas around the theme of Innovating, Integrating, Building – Highly Innovative and Sustainable Buildings. The award ceremony for the winning projects took place at SAIE during the Forum Architecture and the New Generations. The winners in the four areas in the category Young Architects"are:
des + Bettini, Italy, Wood
HArquitectes + dataAE, Spain, Concrete
LAN Architecture, France, Brick
CAFèARCHITETTURA, Italy, Metal&Glass

混凝土结构奖 / PRIZE ARCHITECTS CONCRETE
HArquitectes + dataAE
"UPC Student Housing", Barcelona, Spain
"UPC学生公寓"，巴塞罗那，西班牙

金属与玻璃奖 / PRIZE ARCHITECTS METAL&GLASS
CAFè ARCHITETTURA
"VAL D'OCA Wine Centre", Valdobbiadene, Italy
"VAL D'OCA葡萄酒中心"，瓦尔多比亚代内，意大利

砖结构奖 / PRIZE ARCHITECTS BRICK
LAN Architecture
"RIVP Student residence", Paris, France
"RIVP学生公寓"，巴黎，法国

主编 · DIRECTORS AND PUBLISHERS
Gerardo Mingo Pinacho · Gerardo Mingo Martínez (西)

主办单位 · SPONSORS
西班牙未来建筑 · future arquitecturas s.l.

图形制作 · GRAPHIC PRODUCTION
Gerardo Mingo Martínez (西) · gerardo@arqfuture.com
赵磊 Zhao Lei · leizhao@arqfuture.com
左雯莎 Zuo Wensha · news@arqfuture.com
郑金氏 Zheng Jinshi · communication@arqfuture.com

中国地区公司合伙人 · CORPORATE PARTNER IN CHINA
赵磊 Zhao Lei · leizhao@arqfuture.com
地址 address: 中国杭州下城区文晖路303号
浙江交通集团大厦11楼
邮编 postal code: 310014
电话 telephone: +86 571 85303277
手机 cell phone: +86 13706505166

行政人员 · ADMINISTRATION
Belén Carballedo (西) · belen@arqfuture.com

销售部 · DISTRIBUTION DEPARTMENT
曾江福 Zeng Jiangfu
手机 cell phone: +86 13564489269
电话 telephone: +86 21 65877188

市场推广 · MARKETING DEPARTMENT
曾江河 Zeng Jianghe
手机 cell phone: +86 13564681595
电话 telephone: +86 21 65878760

翻译 · TRANSLATION
王坤 Wang Kun

广告 · ADVERTISING
china@arqfuture.com

future arquitecturas
Rafaela Bonilla 17, 28028
Madrid, Spain

www.arqfuture.com

© 2012 future arquitecturas s.l.

未经未来建筑事先书面授权,本书任何部分都不得被复制、转载、转贴、分发、再发行、销售、修改或者储存在任何一个文件管理系统中,也不得以任何形式或手段对此书进行非法传播。
No part of this publication may be copied, reproduced, reposted, distributed, republished, sold, modified, stored in a document management system, or transmitted in any form or by any means without the prior written consent of future arquitecturas s.l. PANORAMA architecture newspaper is a trademark registered by future arquitecturas s.l.